Louisiana's
Haunted
Plantations

Jill Pascoe

IRONGATE PRESS
Baton Rouge, LA

Photography by Jill Pascoe

ISBN 0-9754746-0-X

Library of Congress Control Number: 2004093375

Published by:
IRONGATE PRESS
P.O. Box 84602
Baton Rouge, LA 70884-4602
www.irongatepress.com
info@irongatepress.com

Cover photograph of Chretien Point Plantation

Cover Design by Doyle Buehler
Printed in Canada

Dedication

For Josh,
who might not believe in ghosts,
but always believes in me.

Table of Contents

Acknowledgements

This undertaking would not have been possible without the enthusiastic support of the plantations featured in this book. I am most grateful to all who shared their stories with me and who helped arrange my visits: Teeta Moss and Hester Eby at The Myrtles Plantation; Polly Luttrull at Rosedown Plantation; Lynette Tanner at Frogmore Plantation; Melinda Fitzgerald Anderson and Beulah Davis at Loyd Hall Plantation; Margaret Belis, Tesa Laviolette, Nell Nations, and Dorothy Theresa White at Chretien Point Plantation; Jane Landry, Sandra Schexnayder, and Trey Castleberry at Oak Alley Plantation; Kim Fontenot, Catherine Stephens, and all the other staff members who helped at San Francisco Plantation; the tour guides at Ormond and Destrehan Plantations; Blossom, Salvador, and Lisa Lentini at La Branche Plantation; Myrna Bergeron at Pitot House; and Foster Creppel and Richard Fern at Woodland Plantation. I would also like to thank my family for their relentless editing efforts without which this final product would never have been possible.

Introduction

Ghost stories have always captured our imagination. Historians began to record these stories as early as the first century A.D. The telling started much earlier.

Shadows, specters, phantoms, spooks, apparitions, revenants, wraiths, and ghosts; we have so many words in the English language to describe something we cannot really explain. We are spellbound by tales of spirits who walk the earth. It may arise from our hunger to understand the unknown, to fathom the ultimate mystery. Stories with ghosts disrupt our everyday reality and sometimes leave a sense of unease. But a little shudder from a ghost story is a unique pleasure. For some it comes from telling a story by the campfire, for others reading by the fireplace at Christmas, and for still more Halloween always conjures up the presence of ghosts.

Ghosts and Louisiana seem to be synonymous. Legends and stories abound about ghostly apparitions and hauntings, especially in New Orleans, a city that many have deemed the most haunted in America. But mystery also lurks in the plantations in this state. There is an old saying in Louisiana that every respectable plantation has at least one ghost. In the stories you will read in the following pages, you will find many spirits lurking in the shadows of these grand homes.

These old plantation homes remind us of a time long since passed. A time when beauty and wealth and privilege for a few came at a terrible price: slavery. Perhaps this violent, tragic, and turbulent past makes Louisiana more haunted than other states. The people who lived and died on these plantations are not all resting in peace. Some of the dead have returned.

Several of the families who lived at the plantations described in these pages experienced turbulent histories. Murder, death through childbirth, disease like the dreaded yellow fever epidemics, dangerous liaisons with pirates, and the Civil War all took people before their time. These unhappy souls along with others who seem to recall joyful times continue to walk the halls of these plantations. And, as we shall see, even plantations burnt to the ground are still haunted by a ghosts or two.

This book places the ghosts within the historical and architectural context of their plantation homes. Louisiana has many glorious plantations with enthralling histories. Here you will find these histories enriched by former residents who still find their homes captivating. The stories that fill these pages are based on readings, research, and by speaking directly with the people who live or work at these plantations.

All of the intact plantations portrayed in this book are on the National Register of Historic Places and with the exception of the burnt plantations, are open to the public for tours or overnight accommodation. So please go and visit these beautiful homes in person. For now, sit back and enjoy your journey through Louisiana's Haunted Plantations.

The Myrtles Plantation

The Myrtles Plantation
America's Most Haunted Home

As you drive up to The Myrtles Plantation, all traces of the present seem to dissolve away behind you. The drive is lined with centuries old live oaks and crepe myrtles draped with Spanish moss like lush green spider webs. Even on a sunny day, there is an eeriness about the plantation, as the house seems always to be cast in twilight. Shadows cascade from overhanging branches through your car windshield as you approach the house. This uneasy feeling only increases as you continue along the curving driveway past weathered stone statues and set eyes on the home before you. The Myrtles Plantation is unlike any other you will read about in this book. Its long galleries and blue-gray lacy ironwork set this regal lady apart from other plantation homes in Louisiana. Around the house there are more trees with moss, and even more stone statues, all of which add an air of mystery to this plantation, called by some the most haunted home in America.

But does The Myrtles deserve this designation? Numerous television programs, newspaper accounts, and journals have said so. The Myrtles has even been declared an authentic haunted house by the United States Department of Tourism and the Smithsonian Institution. With at least seven active spirits roaming the house and the grounds it certainly seems a likely candidate.

General David Bradford built The Myrtles Plantation in 1796. General Bradford was known as "Whiskey Dave" of the Pennsylvania Whiskey Rebellion. Western Pennsylvania residents, who were already agitating to separate from the United States, became particularly incensed by a tariff placed on the sale of whiskey in 1794. This tax was collected at the source instead of the point of sale. Western whiskey from Pennsylvania was taxed at twenty-eight percent and eastern whiskey was taxed at only fourteen percent. A group of rebels formed and in July 1794, Bradford assumed leadership. George Washington began to mobilize his troops in August of the same year and by October they began to round up the rebels. David Bradford escaped and fled to Louisiana to avoid being imprisoned. Here he acquired six hundred and fifty acres of land through a Spanish land grant. He built his home as a simple four-room cottage on land rumored to be sacred Tunica Indian burial ground.

The home standing today reflects the changes and additions made by the Stirling family in 1834. They increased the size of the property to five thousand acres and added the entire south wing of the home including the formal foyer, staircase, dormers, and double dormers. Elaborate details were added to the house including the Rococo Revival open pierced plaster frieze work and medallions, the stunning baccarat crystal chandeliers, impressive Carrara marble mantelpieces, and the grapevine patterned cast-iron work. This cast-iron work on the galleries of the house is unusual in plantation architecture in Louisiana, being much more prevalent in the townhouses of New Orleans. In the main floor of the house are matching lady's and gentleman's parlors, a dining room, a gaming room, a lady's day bedchamber with gold leafed French furnishings, and another bedroom suite used for bed and breakfast guests. The first floor also boasts floor to ceiling windows and French doors, including hand-painted glass entrance doors to the foyer. The second storey of the house has five bedrooms for bed and breakfast guests. Today the home is beautifully restored and available for tours, as a bed and breakfast, and for special events. As an added bonus The Myrtles retains much of its original furnishings.

Despite the beauty of the home, darkness always seems to engulf it. No wonder as this home has witnessed many tragedies in its over two hundred year history. The story most often retold is about Chloe and the oleander cake. In 1818 The Myrtles passed to Sarah Matilda (General Bradford's daughter) and her husband Judge Clarke Woodruff. While living in the home they owned a house slave named Chloe. She was apparently very curious and wanted to know the affairs of the house. She is also said to have been Judge Woodruff's mistress, a common practice at the time. In the early 1820s Chloe was caught eavesdropping outside either the gentleman's parlor or the dining room by Judge Clark Woodruff. In his fury over Chloe's misbehavior Judge Woodruff ordered her left ear cut off. She was also banished from work in the house and forced to a life of labor in the fields. Afterwards Chloe wore a green turban around her head to cover the disfigurement.

In an attempt to make amends to the family and regain her position in the house Chloe baked a birthday cake for one of the Judge's daughters. However, this was no ordinary birthday cake. It contained oleander leaves, which produce a poison like arsenic. Some believe Chloe did not wish to kill anyone in the family, but merely make them sick. What's more, since she knew the cause of their ailment, she could cure them and regain her stature in the house. It is of course possible that she baked the cake as an act of terrible revenge to the Judge, but The Myrtles' staff prefer to think she had good intentions. Chloe's plan failed miserably. The women of the family fell deathly ill and Chloe could not cure them. A terrified Chloe confessed to the other slaves in the hope that someone wiser than she would be able to cure the ailing family, but it was too late. Sarah Matilda and her two daughters were dead by daybreak. In their horror over what she had done, the other slaves hanged Chloe and threw her corpse into the river.

This single event accounts for four of the murders at The Myrtles Plantation and much of the haunting as well. Sarah Matilda is a quiet spirit, occasionally heard softly crying, but the little girls and Chloe are active all over the house. They are generally seen in the older section of the house, where they lived, but they have also ventured into the "new" wing. The children spend a lot of time in the Ruffin-Stirling Room, which was their nursery. Guests have seen the children in this room, close to the fireplace, reliving their final hours of anguish, convulsed with abdominal pains, vomiting, careening from dizziness, and finally collapsing dead as their overstressed hearts stopped. Others have heard only crying and wailing in the room. On numerous occasions overnight guests in an adjacent room will ask the plantation staff if the child staying in the Ruffin-Stirling room is okay, because the little one sounded so ill during the night. Of course there was no child in the room and often no adult guests either.

The children also seem fond of playing in the Fannie Williamson room, where dolls are known to appear and disappear. Originally six dolls graced the mantle, but Manager Hester Eby insists there is only one now. However, on occasion guests will say all of the dolls in the room are very pretty. When questioned they say there are as many as six dolls. Could this be evidence of the children playing? There have been sightings of little girls peeking through the glass doors into the lady's parlor. During their lifetime the formal parlors were likely restricted from children, so now they still seemed trapped outside but perhaps bold enough to peer into the room. Although the south wing of the house was built after the deaths of the little girls, they do seem interested in the General David Bradford Suite where people have experienced someone snuggling up in bed with them.

Chloe herself has been spotted by guests at The Myrtles and even appeared in a photograph taken by owner Teeta Moss. Guests have reported a sense of dread as if someone was watching over them while in bed. When they open their eyes, a woman with a green turban is peering intently at them. Sometimes guests report being tucked tightly into bed as if they were children.

At the time of their murders, a commonplace belief was that a person's soul could be trapped in a mirror, and thus were shrouded with black fabric when someone died. However, in the chaos and confusion surrounding the deaths of the girls and Sarah, one mirror was never covered and their souls are believed to have entered it. Strange streaks and handprints appeared on the glass almost immediately. Over the years the glass has been replaced often, only to have the same strange features return. Hester Eby, who has worked at The Myrtles for fifteen years, has seen the glass changed eight or nine times always to have the irregularities return. The last time the glass was replaced was shortly after the current owners took over the house in

December 1992. Within two weeks the markings suddenly returned. While past owners changed the glass repeatedly to see if the phenomenon reoccurred, the current owners have no such plans. The mother and daughters have pressed their ghostly hands and faces against too many mirrors already.

The souls of the girls and their mother trapped in the mirror could be responsible for the strange photographs that visitors have taken. During the tour of The Myrtles guests can take pictures only of the mirror, and often strange and unexplained things will be revealed. During my stay at The Myrtles in July 2003 I took two photographs of this mirror from different angles. Both photographs show images that are difficult to explain. In the upper right hand side of one photograph, there appears to be a floating skull, and in the other what appears to be the profile head of a jester. Could these images be the result of the spirits of Sarah Matilda and her children?

Tragedies continued to befall The Myrtles Plantation. Ruffin Stirling and his wife Mary Catherine bought the property in 1834 and did extensive renovations and an expansion of the house. However, aware that the house was already haunted, they took measures to protect themselves from their otherworldly visitors. Acanthus leaves (a symbol of protection) were carved into the elaborate pierced plaster frieze work. In the lady's day bedchamber there are cherubs looking down into the corners of the room from the chandelier to ward off intruders and there are the heads of nuns in the medallions. The locks were placed upside down to confuse the ghosts and covers were placed over the locks as added protection against supernatural house guests. Stained glass windows at the front and back porches have hand-painted crosses as another measure to ward off evil spirits. Even though the Stirlings took all these precautions tragedy would befall this family as well.

The Stirlings had a large family with eight sons and one daughter. Several of the sons went off to fight in the Civil War, never to return to The Myrtles. Their eldest son Louis was killed over a gambling debt on the first floor of the house in the gaming room; it is said that his spirit still haunts the house. It is believed one son did manage to survive. A tutor was also shot dead during the Civil War in the gentleman's parlor. The house eventually passed along to Sara, the Stirlings only daughter, who married an attorney, William Winters. They had one daughter Kate, who tragically caught yellow fever in 1861 at the age of three. Supposedly a voodoo priestess named Cleo was called in to help save the child, but she was not successful and Kate died. Kate's sad spirit can still be felt in the room where she died. Legend has it the voodoo priestess also haunts The Myrtles where she is forever doomed to fail trying to save a young life.

As if Sara had not experienced enough tragedies in her life, one more followed. In 1871 a rider arrived at the house asking to see her husband, the lawyer. As soon as William Winters stepped out onto the porch he was shot. He managed to make his way back into the house and climb to the seventeenth step where Sara met him and he died in her arms. This event sent Sara into mourning for the rest of her life, shutting herself in the house and spending most of her time in a rocking chair in their bedroom. Over the years visitors have reported hearing the ghostly stumbling steps of William Winters as he struggles up to the seventeenth step, but no further. Sara can also be seen and heard in the Judge Clark Woodruff suite. Visitors have heard a chair rocking, even though there is no rocking chair in the room, and been engulfed in the scent of a floral perfume. While staying in the suite on the anniversary of Sara's birthday I experienced a sudden intense floral scent that just as quickly evaporated. This happened before I learned a fleeting floral aroma is typical of this room and that no floral cleaning agents of any kind are used at The Myrtles.

One more ghost haunts The Myrtles Plantation. During 1927 an overseer of the plantation was shot during an attempted robbery. This spirit, dressed in overalls, has been spotted at the entry gate telling people to go home as the plantation is closed. He also manifests himself at the overseer's cottage, which is now used for bed and breakfast guests. When photographs are taken his face often appears looking out of the windows. Overnight guests in the cottage have also heard the sounds of someone making breakfast in the morning, only to find no one there.

Like every plantation with several spirits on the loose, strange and unexplained things seem to happen all the time at The Myrtles. Footfalls from unearthly sources are heard, doors open and close of their own will. Names are called by thin, other worldly voices, most often in the foyer. There is the heady scent of honeysuckle when it is not in season; the aroma of freshly shucked corn when none is to be found. Every Friday and Saturday night staff at The Myrtles Plantation give mystery tours of the house. As you wander through the rooms the haunted history is told. To set the mood for the tour, staff dim the lights, but during the tour unseen little hands often adjust the lights up and down. The result of ghostly children playing? Mounds of sand have been found in the house along with little wet footprints on the carpet. Many guests report hearing the sound of music and glasses tinkling, like a party is happening downstairs in the house. When they question the staff the next day, they find out that part of the house was locked and empty the previous night.

Once when Hester Eby was outside with the owner of a tour company they heard a loud crash from inside the house. Hester describes what the noise was like, "I thought the chandelier had fallen in the middle of the dining room table and all the china and everything was broken. It just really sounded like that. Like the table was just going to be caved in." On inspecting the room, everything was in place, nothing was broken, and no earthly source for the sound was found. Both women just looked at each other and silently left the house.

Every room at The Myrtles has a story to tell and ghostly occurrences can happen at any time. Keep the following in mind when making a reservation. But do remember, payment is required up front, so do not leave in the middle of the night, and hauntings are not guaranteed.

William Winters Room~ guests in this room often wake up to find themselves well tucked into bed, from head to foot. This is also a room that has been described as feeling crowded, as if you are not alone. Staff do not like spending time in this room.

Ruffin-Stirling Room~ the former nursery is where overnight guests report the sounds of children crying, forever reliving their agonizing deaths.

John Leake Room~ sometimes the smell of cigar smoke is strong in this room. It is believed that a confederate soldier died in this room. Guests have reported seeing a confederate uniform, sometimes hanging in the armoire and other times on the bed. The Myrtles Plantation owns no such uniform.

Fannie Williamson Room~ in this room items are rearranged or go missing, like the dolls, perhaps the result of the girls playing. A producer from the Oprah Winfrey Show had a terrifying experience in this room. She was tired and resting on the bed. Suddenly the room became freezing cold. Shivering, she decided to go back downstairs. She tried the door but it wouldn't open. It seemed stuck or locked or barred. She kept pulling at the door, becoming increasingly afraid in the freezing room. Finally the door gave way but as she rushed out she felt icy fingers slipping from her shoulder. She escaped down the stairs very quickly. The crew cut their visit short.

General David Bradford Suite~ this is the room where guests report the feeling of someone getting into bed with them and snuggling up beside them.

Judge Clark Woodruff Suite~ the final footfalls of William Winter on the stairs are heard here, the scent from roses is sometimes overwhelming, and the hallway often feels very cold, even during the steamy Louisiana summer.

The Caretaker's Cottage~ the caretaker who was shot in 1927 has apparently never truly left his job as he is seen frequently in this, his former home.

Garden Rooms (Azalea, Camellia, Magnolia, and Oleander Rooms) ~ the carriage house burnt down sometime in the 1800s leaving behind only the flooring, which is currently visible in the OxBow Carriage House Restaurant. Adjacent to the restaurant are four garden rooms. Some guests have reported they hear screams and fingernails on wood like someone is struggling to get out of the room. One guest dreamt about horses burning in a fire before knowing about the carriage house fire. Hester Eby comments there was a time, before the plantation was so busy, that you could feel the earth moving as if a herd of animals, perhaps horses, was thundering towards you.

With at least seven ghosts and more murders and tragedies than any home should have witnessed The Myrtles Plantation is indeed one of the most haunted houses in America. You can see for yourself and stay in the house or visit the grounds, but do be brave and make it through the night. For the faint of heart, the house always seems much more peaceful during the day.

2

Rosedown Plantation
An Impish Ghost

W hen the subject of ghosts in St. Francisville is raised, thoughts naturally turn to The Myrtles Plantation. But it is not the only haunted plantation in St. Francisville. Rosedown Plantation also has a resident ghost that wanders the house, grounds, and gardens of this beautiful estate.

Daniel and Martha Turnbull built Rosedown Plantation between November 1834 and May 1835. It was constructed by Wendall Wright with double galleries and a neoclassical Doric columned façade. The wings of the house, added in 1845, resemble small Greek temples. Under the management of the

Turnbulls the plantation swelled in size to 3,455 acres, the majority with cotton fields. Besides the beautiful Classical style house and rich earth planted with cotton, Rosedown, named after a play the Turnbulls saw in Europe during their honeymoon, was and still is, known for its extensive gardens. It was also during their honeymoon that they viewed the great formal gardens of Italy and France and Martha Turnbull devised her own ideas for the gardens on her future plantation. Eventually her gardens would cover twenty-eight acres, all of which remain today, restored to their former beauty.

Like most plantation homes Rosedown produced many of the materials required for construction on-site. Cypress and cedar trees from their own swamp and processed at their own sawmill provided the necessary wood for the majority of the structure of the house. The delicately fluted columns on the façade were carved at their sawmill as well. More elaborate and rare materials, such as the marble fireplace mantles and the seventy feet of mahogany for the stairwell banister, were imported either from the northern states or from abroad. The house has beautiful cornices and medallions in the first floor rooms and an elaborate entrance foyer with a sweeping spiral mahogany staircase. The downstairs has the formal rooms of the house in keeping with the Classical style, diverging from the Louisiana French Colonial style seen at other plantations. Formal rooms include a parlor, dining room, music room, gaming room, and breakfast room. The second floor of the house has the bedrooms and also a large central hallway blessed with cooling breezes off the front balcony.

Rosedown is unique in that the wings of the home were not added for more space for a growing family, as is usually the case, but rather to accommodate an unusual set of furniture. Henry Clay ran unsuccessfully for president of the United States in 1824 and then ran again in 1844. In anticipation of a win some of his friends engaged Crawford Riddle 'N Journeymen Cabinetmakers in Philadelphia, to create a Gothic bedroom suite for his use once he ascended into the White House. However, Clay's ambiguous stand on slavery cost him the election and his friends decided to sell the furniture. Daniel Turnbull bought this bedroom set for Rosedown, but because of its large size he needed to build a wing onto the house. This was the north wing; the south wing was added to create the necessary balance and was used as a library. So, a lost presidential election led to the two wings of Rosedown. Sadly this set of furniture was sold prior to the State of Louisiana acquiring the property. Rosedown does however retain much of its original furnishings, and has delicate touches such as unique curtain tiebacks in each room, one set resembles arms and hands.

Descendents of Martha and Daniel Turnbull managed to keep Rosedown Plantation in family hands until 1956 without a mortgage, but lacked the funds to maintain and restore the house and tend to the gardens. When Catherine Fonden Underwood purchased the property in 1956, the house needed extensive repairs and the gardens were completely overgrown. Yet both echoed a sense of their former beauty. Mrs. Underwood restored the house and gardens at Rosedown and opened the house as a museum. Martha Turnbull's own garden journal was instrumental in this restoration.

It is possible to spend an entire afternoon in the gardens at Rosedown; from the alley of live oaks draped in Spanish moss, to the various summerhouses, arboretums, statues, fountains, and flower gardens, the entire property is enchanted. But as you roam the grounds, a figure you see out of the corner of your eye might not be a modern visitor at all, but a shadowy reminder of Rosedown's past.

Rosedown Plantation

It is believed the ghost who haunts Rosedown plantation is William B. Turnbull, eldest son of Martha and Daniel, who drowned in 1856 at the age of twenty-seven. He went under while attempting to cross Old River at his DeSoto Island

Plantation when the wake of the steamboat Bella Donna upset his skiff. Since hauntings are often associated with violent deaths, or strong emotions, it seems likely to the staff at Rosedown that their resident ghost is William Turnbull. Adding credence to this thought is a permanent reminder of Martha and Daniel's grief at losing their son. On the underside of the roof of the south wing are black mourning stripes that forever memorialize the death of William.

The ghost at Rosedown loves to play jokes with the staff like a young prankster. William's antics at Rosedown happen everywhere on the property, although he seems particularly fond of tinkering with the lights in the house. While Curator Polly Luttrull confessed her amazement that he could understand the complex electrical system in the main house, he has had a long time to figure it out. Ms. Luttrull describes her first experience with the lights as follows: "I just finished the last tour and I had gone through the house to shut off all the lights prior to locking up and leaving. After I shut off the last light in the library and came back through the house every light was on in the house again. I had to go back and turn them back off. It gave me a very eerie feeling when it happened."

Ms. Luttrull is not the only employee to experience strange happenings with the lights at Rosedown. Sometimes the whole house will be lit up and other times only a few rooms. Ms. Luttrull describes a particularly trying experience that occurred to a couple of tour guides one night. "They said it was weird one day. They had turned off all the lights, they had set the alarm, locked the house, and started to walk out to the parking lot, [one of them] said she turned back to look at the house and all the lights were back on. So they had to go back, unlock the door, turn off the alarm, turn off the lights, and re-alarm the house. She said that happened two times in succession that night; they had to turn off all the lights again twice."

Was William trying to get the attention of these women with his ghostly antics? In this instance I am sure these two were glad they were not alone. Not that William ever gives a sense of dread or harm. He just seems to want attention, like a soul who wants to be a child again and enjoy the games and fun of youth.

Another incident with the lights was odd. One morning a tour guide arrived to find the house brightly lit (except for the staircase and formal parlor) although all the lights had been turned off the night before and the house was still armed. Also unusual was that the fountain had been turned off (its switch is also in the main house) which had never previously happened. Despite all of the lighting irregularities that occur, the alarm has never been set off. After all, ghosts do not seem to bother modern alarms and the alarms do not notice and report on ghosts either.

Curator Luttrull has been distressed by one other strange phenomenon at the house. One day when she arrived, a staff member hurried her along to show something odd in the north wing. "We noticed the dressing bureau had shifted out of its normal space. It wasn't anything dramatic, but it had been

28

pushed back against the window and canted slightly, about four inches. And there was dried mud on the floor in front of it, and I know there wasn't anything like that when I closed down the house the night before." Somehow during the night this bureau, part of the furniture bought by the Turnbulls in 1835, had inexplicably moved. To add to the mystery the alarm was still set when the first worker arrived that morning. I examined the bureau, set in its proper location. There were several fragile knickknacks and ornaments arranged on the bureau top and when I questioned Ms. Luttrull she stated that they had all been in their proper places when they found the bureau moved. This bureau is a large, substantial piece of furniture, which would take more than one person to move without leaving marks on the hardwood floors underneath. Yet there were no scratches on the gleaming hardwood floors. There was only the dried mud right in front but nowhere else leaving no earthly source as a likely culprit. Could this prank be the work of William, or is it the result of another spirit?

William does seem shy about letting himself be seen at Rosedown. Over the years he has only shown himself to one staff member, a former tour guide. This guide once glimpsed William off to the side moving across the plantation grounds, but it was so fleeting that when he turned there was no one in sight. He was certain however that what he saw was the figure of a man.

On another occasion, a couple visiting from Texas had an unusual experience while looking out at the gardens from the main porch and taking some pictures. While the husband photographed his wife, she felt a strange, sudden, and unnatural chill. She described it as a wave of cold passing through her. It seemed so cold she hurried into the gift shop to warm up. When the pictures were developed the couple was shocked to see unexplained anomalies in two of the images on their film. These two photographs were the ones taken when the woman felt so cold. There appears to be the arm and neck of someone other than the woman in the pictures and the face of the woman is blurred like something was moving through her, while the rest of the photograph is perfectly clear. Could this be William experimenting with photography? We can imagine William vanishing with a sly, impish smile just before being seen. He is a supernatural who works through the modern magic of electricity to confound us. During my visit I photographed extensively in the marvelous gardens and no strange anomalies appeared in my pictures. Perhaps William, knowing my purpose, remained shy that day.

Recently Rosedown Plantation became part of the Louisiana State Park System, thus ensuring its preservation. Visiting Rosedown is a feast for the senses. From the beauty of the house, to the extensiveness of the outbuildings, to the lushness and rich scent of the gardens, this plantation is worth an extended stay. And perhaps late one evening as you linger over tea, William will turn off your lamps, sending a chill down your fingers.

3

Frogmore Plantation & Gins
Elusive Ghosts

The first thing you notice when you approach Frogmore Plantation is the cotton. Lush fields of white surround the historic buildings, which dot the landscape. Frogmore is located in northeastern Louisiana, only seventeen miles from Natchez, Mississippi. The plantation home is a one and a half story raised cottage and is still standing after nearly two hundred years. Beyond the plantation are Indian mounds, which like the historic buildings on the property, are on the National Register of Historic Places.

At Frogmore, you can see technology in all its glory in a modern cotton gin and also be amazed at how advanced they were one hundred and twenty years ago with a historic gin. What's more as you explore the restored grounds and buildings, you are transported back in time to a living and working Louisiana plantation of old. Such is the beauty of Frogmore. The nearness of the past is felt in other ways as well, due to the presence of three ghosts.

Daniel Morris acquired Frogmore Plantation and built his house in 1815. It is believed that this home began as a raised dogtrot house, comprised of only three rooms and raised a full story above the ground. However, when Daniel married in 1832 his wife brought two children into the home from a previous marriage (her first husband had died). With the sudden increase in his family, Daniel expanded the house, adding the cabinets (back bedrooms), the loggia, and a half story to the house. The front of the house has a gallery supported by Tuscan columns. There is a double-pitched roof and the chimneys are on the end-walls, which was not typical of Creole houses. It is uncertain how much acreage Daniel Morris had when he operated the plantation, as land records were not officially recorded until 1843. He is thought to have owned four to five hundred acres.

In addition to the main house, Frogmore Plantation has a number of dependencies (outbuildings to the big plantation house); some are original to the plantation while others were moved from neighboring plantations. Included in these outbuildings are a commissary, a dogtrot cottage (used as an overseer's cottage), a cooking cabin, a church, a washhouse, a pigeonnier, a smokehouse, a three-hole privy, a planter's office, and eight slave quarters (one of which is original to the plantation). The buildings moved to Frogmore were placed as close as possible to the location of the original buildings.

The plantation houses an 1884 Munger steam gin, which was moved to the site from Rodney, Mississippi. Frogmore has run a public gin since the 1800s and still operates a modern public gin for the surrounding producers. While many plantations produced cotton, they could not all afford to have gins, so they would take their cotton to Frogmore.

1884 Munger steam gin, Frogmore Plantation

Tragically, Daniel Morris' wife died after giving birth to their son in 1836. Unable to care for three children, including a newborn son, Daniel's two stepchildren went to live with his wife's brother. In 1839 Daniel died and his son went to Natchez, Mississippi to live with his mother's sister-in-law from her first marriage.

After Daniel Morris died, John and Susie Gillespie purchased the property. The Gillespies enlarged the plantation to two thousand six hundred and forty acres. John Gillespie was a very wealthy man and owned eight other plantations. They had five children, but sadly John and Susie were killed in 1855 in a train wreck in Pennsylvania. As a result their twenty-one year old son, William, inherited the property. During the Civil War, Frogmore was strategically located between Natchez and Trinity and a battle was even fought in front of the house. William retained the property until 1867, when the troubles of the Civil War, lack of slave labor, and other economic hardships forced him to sell the property to a wealthy New Orleans cotton merchant named Weis.

Weis hired Benjamin Wade to oversee Frogmore Plantation. Benjamin's family owned Prospect Plantation near Natchez. When Benjamin Wade's father died in 1902 Benjamin's brother stayed at Prospect Hill Plantation, and Benjamin, after thirty years of working at Frogmore bought the plantation.

Benjamin Wade married his first cousin and they had seven children, of whom only two survived to adulthood. After the death of the parents the property was divided between the two children, a son and a daughter named Anna. Anna retained her portion of the property, which included the plantation house, until her death in 1984 at the wonderful age of 101. Her brother's portion was divided between many different owners when he died without heirs. Anna never had children, apparently because she was afraid after losing so many siblings. She also never wanted to live in the plantation home, feeling it held bad memories, and let the overseers of the plantation use it as their quarters, while she lived a short distance away.

Buddy Tanner began leasing Frogmore in 1964-1965, and the Tanners bought the property from the heirs of Anna in 1984. Since acquiring Anna's one thousand three hundred and twenty acres, Buddy and his wife Lynette have been able to buy two other parcels of land bringing the acreage of the plantation up to eighteen hundred acres. They operate the plantation as a working cotton plantation with a modern gin and conduct both historical and modern tours of the property. They use the slave narratives, collected by the Workers Progress Administration (WPA) in the 1930s, to furnish a thorough and accurate portrayal of slave culture and the plantation system, including a tour that reenacts a slave wedding. Visitors can also experience first hand how to pick cotton.

During the years Anna owned Frogmore Plantation several different families lived in the plantation house as overseer of the property. The Tanners have undertaken an extensive oral history project with former workers and tenants of the plantation grounds. This is a brilliant idea and makes their tours as informative as possible and ensures the preservation of the history of the plantation. One thing all these families seem to have in common is visitations from beyond; Frogmore Plantation appears to be haunted.

Mack Jefferson and his brother worked as sharecroppers at Frogmore in the early 1900s and then as gardeners after the end of the sharecropping system. They put a horseshoe on the front door of the plantation house to keep it free of ghosts, but it must not have worked as they insisted that the home was haunted.

In the 1940s, the Sojourners lived in the house along with the ghosts. Once Mrs. Boyd Sojourner saw a man dressed in a white suit walk across the hallway from a bedroom toward what is now a parlor. He then disappeared. She clearly saw him walk across the twelve-foot wide hallway before his figure dissolved before her eyes.

Around 1951 Jack Ellard and his family were living in the main house and one day he was standing outside the house with his daughter. He and the little girl saw a woman, all dressed in black with a veil, standing pensively by one of the columns of the house. He told his child not be frightened as it was her mother. As he was uttering these reassuring words, the woman in black abruptly vanished. He ran into the house where his wife was sitting and asked her why she had scared them like that. She insisted she hadn't been outside.

The mysterious woman in black was not the only encounter with the unknown Mr. Ellard experienced. Regularly, the lights in the bedroom flashed on and off by themselves. He heard chains rattling in one of the back bedrooms, which was always kept locked. On three occasions he saw the front door knob turn and the door swing open, letting in only the night. This door has the original carpenter lock that dates to the building of the house in 1815. It is operated by a long skeleton key and remains the only way to lock the house to this day. Another time the clock strangely rang at three o'clock. Normally, this would not be an odd occurrence, except that the clock had not been wound in years (Mr. Ellard speculates it had been at least fifty years since it was wound), and the clock actually struck the correct hour. Since this bizarre occurrence the clock has remained silent. Despite sharing his home with ghosts Mr. Ellard took comfort in the fact that they were friendly ghosts and they did not seem to bother his family except for some small unexplained events. Everyday reality was disrupted just slightly by these ghosts as they ventured away from their eternal abode.

Frogmore Plantation

When Lynette Tanner first moved into the house she constantly heard heavy footsteps on the wooden stairs to the second floor. She would run up to make sure her children were okay and they were always sound asleep; no source for the footsteps was ever found. This occurred many times over two years. In 1993 the Tanners completely restored and redecorated the house. The two large storage rooms under the house were enclosed and turned into bedrooms, a bathroom, and a game room. After the restoration of the house, the ghostly footsteps ceased completely. Perhaps the spirit has finally reached his destination in the beyond and no longer suffers the burden of climbing the stairs endlessly. Lynette has not experienced any other unexplained events. In fact, no other ghostly events have occurred at all in the house since the restoration, except an apparition seen by Lynette's high school aged daughter in 1999.

Late one night Lynette's daughter was startled awake. Vaguely at first but then clearly she saw a house slave walking through her bedroom door. The apparition had not opened the door but was actually passing slowly through the solid door. The house slave was dressed in dark clothing with a red apron. This was disquieting as the woman seemed real, as if she was alive but yet unearthly. Lynette's daughter had to turn away not believing what she saw. When she dared to look again the ghost was still there. The girl hollered for help but by the time Lynette arrived the image had faded away.

The bedrooms where Lynette and her daughter were

sleeping were in the portion of the house that had been two open storage rooms, until being enclosed in 1993. While the ghosts who formerly appeared in the main house seem satisfied with the Tanners' restoration work and have not yet returned, perhaps this phantom was confused as to why her former work area was now occupied. Since the appearance of this ghost all has been quiet at Frogmore Plantation.

Questions still remain about the wraithlike visitors. Were the ghosts of the woman and the man seen at the house in the 1940s and 1950s Daniel Morris and his wife roaming the grounds of their house, reliving their tragic lives? The exact identity of the house slave is also unknown. These three spirits seem to be at peace. For now Frogmore Plantation no longer seems haunted, but you never can be certain what will raise the ghosts again. In the future it is possible a restless spirit will return, perhaps the sad woman seeking the comfort and solitude of the columns or the spirit trudging up the stairs bearing the weight of the ages.

Loyd Hall Plantation

4

Loyd Hall Plantation
A Mischievous Spook

A devilish plantation owner hanged on his own grounds, a tragic young girl who killed herself after being abandoned at the altar, a poisoned house slave, and a love-lorn Union soldier shot dead during a struggle for a gun. These four restless spirits haunt the house and grounds of Loyd Hall Plantation.

William Loyd built Loyd Hall Plantation in 1820. He was the black sheep of the Lloyd insurance family of London and reportedly was paid a large sum of money to leave England and never return. Part of the deal included William changing the spelling of his family name from Lloyd to Loyd. Upon his arrival in the United States, William journeyed to the Carolina/

Tennessee area where he met and married his wife Sarah. The newlyweds traveled to Louisiana via the Mississippi River, the Red River, and Bayou Boeuf. Eventually they purchased six hundred and forty acres in the area of Cheneyville and started building Loyd Hall Plantation. The plantation had its own lumber and brick mills, and with the labor of their sixty slaves, the Loyds built the big plantation home that stands to this day. The land was planted with sugarcane, cotton, tobacco, and indigo.

The walls of the home are sixteen inches thick and go three feet into the ground, a construction technique that has helped preserve the house. Loyd Hall is two and a half stories with paired end chimneys and a central hall plan. This is classic Georgian architecture, although the galleries and balconies signal a home designed for the Southern climate. Later additions to the home circa 1850 include the ornate plaster work throughout the first floor of the home; the plasterwork is particularly stunning in the parlor, dining room, and entry hall. The cast iron balustrades and decorative entablatures over the doorways were also added at that time. In 1948 the rear double wooden galleries were enclosed to allow for indoor plumbing in the house. Ceiling heights are impressive on all levels of the house with sixteen feet on the first floor, fourteen feet on the second, and twelve in the attic. The large windows, high ceilings, and transoms present throughout the house help to moderate the intense heat of Louisiana's summers.

After William Loyd died in 1864, Mrs. Loyd and her two

sons remained in the house until 1871. Very little is known about the house between the years 1871 and 1948 when the house changed hands twenty times. It's not clear what caused this rapid turnover, but we can suspect the ghosts. The over seventy-five years of neglect were not good for the house. In 1948 the Fitzgerald family purchased the property and they still live on it today, although five years ago they moved out of the main house and built a home next door. When the Fitzgeralds purchased the land they did not even realize they were acquiring a plantation home. They were interested in finding some farming land and Loyd Hall Plantation was on the market with all of its original six hundred and forty acres. As they flew over the property to inspect the land they did not notice the plantation house, only what appeared to be deteriorated outbuildings. After decades of neglect the home had become so overgrown it could not be seen from the plane.

Despite the decay of the years the Fitzgeralds had unknowingly purchased a gem. They were delighted to discover the plantation house and lovingly restored the damaged Georgian hall to its former grandeur, including its prized plasterwork and suspended staircase. Today the home is beautifully restored and is open for tours and special events.

While the generations of Fitzgeralds resided in the house they learned that more than the beautiful plasterwork and millwork remained of Loyd Hall's storied history; four phantoms still inhabited the home as well.

The first ghostly resident of Loyd Hall Plantation is

Mr. William Loyd himself. During his life Mr. Loyd seemed to stir up trouble wherever he went, a character flaw that no doubt caused his banishment from England. He instigated enough disputes with the local Choctaw tribe that they attacked his house. Two arrowheads can still be seen on what was the front door of the house (the door has now been moved to the dining room). The Loyds survived this attack but William was not so lucky with his double-dealings during the Civil War. Somehow he worked as a spy for both the Confederate and Union armies. This counterespionage did not last long. Union Soldiers uncovered his betrayal and hanged him from a tree on the plantation property. William did not pass on quietly and is thought to be responsible for most of the eerie activity in the house.

The Loyds' grand plantation was a perfect setting for family celebrations. On one occasion, William Loyd's niece, Inez Loyd, was to be married in the house. We can imagine the house decorated with flowers and overflowing with wedding guests dressed in their finest to celebrate the happy day. But the bridegroom did not arrive, abandoning Inez. Grief and shame darkened her spirit as she waited in despair for her love to arrive. Finally, distraught and inconsolable she threw herself through a third story window, crashing down in broken glass to her death.

During the Civil War, Union troops briefly occupied the house and when they decamped one young soldier decided to stay. It is believed Harry Henry deserted because he fell in love with one of the young women residing in the house. To be near his love and out of sight, he hid in the attic. But this was not a safe spot. Harry was discovered in the attic by another female resident of the house. There was a struggle over the gun he kept for protection and it discharged, killing him instantly. His body was buried underneath the house in a shallow grave. Evidence of this tragic death is still visible in the attic where a large dark stain spreads like blood across the unfinished floors.

Sally Boston is the fourth spirit who still drifts wistfully through the rooms of Loyd Hall Plantation. She was a slave nanny who lived in the house with the Loyd family. The exact cause of death is shrouded in mystery, but it is thought she was poisoned. Whatever happened she has refused to leave the home in which she worked.

Chief Hostess Beulah Davis has worked at Loyd Hall Plantation for thirty-two years, starting as a housekeeper for the Fitzgerald family. She is well acquainted with the disturbances of the various resident ghosts of Loyd Hall plantation. While Ms. Davis readily admits she believes William Loyd is responsible for most of the strange occurrences in the house, she has encountered other spirits and has heard many tales of the apparitions from visitors.

When Ms. Davis first came to work at Loyd Hall Plantation, she assumed the odd activity was simply the result of an overactive imagination and being alone in a big old creaking house. However, the frequency of apparitions and strange events over a thirty-two year period rules out simple fantasy. Ms. Davis summarizes the peculiar events at Loyd Hall attributed to William Loyd: "He was very mischievous, he could stir up trouble, and if he could spook you in any way I think he would be here, because he played with your mind. There are times in the house when you can literally hear footsteps very clear, doors open and close...you can walk into some of the rooms and you can feel the presence very strong, sometimes a cold chill will be felt in certain spots. You break out with a cold chill, like goose bumps. And then sometimes it's warm and steamy like a warm breath on your neck and your hands and arms. And sometimes you feel a touch on your skin and you look around and there is nobody there but you can feel the touch." William Loyd is also thought to be responsible for pilfering silverware, napkins, and glasses from fully set tables. Ms. Davis is certain staff did not just forget to set certain pieces as the missing tableware is found later in odd locations, such as bedrooms. William likes to play jokes. Sometimes invisible fingers sound thinly on the piano, the doorbell will ring and no one will be there or the heavy front door will open of its own will. William also seems to like the older form of recorded music. The stereo will abruptly stop and the antique crank-type phonograph will start playing, yet no visible person is near.

Beulah Davis has seen a spirit she believes is Inez Loyd in the house. She describes this encounter: "I saw a tall slender woman with long brown hair and a black dress. I was standing in that door at the end of the hallway coming in [from the back porch] and I was looking into the kitchen doorway. I was standing on the back porch enough to see the secretary sitting at the desk and she had a black dress on and was tall and slender. I am looking at her, and yet going through the kitchen was a tall slender woman with long brown hair….and it was just like she disappeared into the woodwork. She just disappeared, one minute she was there and one minute she was gone." On three occasions Ms. Davis has distinctly heard her own name "Beulah" being called by a woman's voice when no one was in the house. She believes either Inez Loyd or Sally Boston was calling for her but she does not know what they wanted.

When the second generation of Fitzgeralds moved to Loyd Hall Plantation with three young girls their mother made a playroom for them in the attic on the third floor. After playtime, the girls would always talk about their friend "Harry." They described him as being older, perhaps in his early twenties with dark hair, in a Union soldier's uniform. It was only later while attending college, the girls learned of Harry Henry, the young soldier shot to death in their attic. But Harry was never a malevolent specter. On the contrary, he was very protective of the girls as they grew up and seemed to watch over them. One of the girls had a kidney removed and needed a lot of bed rest. Harry would sit at the foot of the bed and watch over her to make sure she fully recovered. These days, Harry can still be heard sometimes pacing the attic floor. At midnight when there's a full moon it is said that Harry Henry appears on the upstairs balcony and plays the violin, lamenting his lost love.

The ghost of Sally Boston is also active in the house and the grounds. There are times when people in the house have caught a fleeting glimpse of a black woman dressed in white. When this specter is seen there is an overwhelming feeling of comfort and an aroma of food cooking. Sally also seems to manifest herself in the back parlor. In this room it is impossible to keep tall tapered candles standing on the mantle; they always end up on the floor. One of the ladies who worked in the house once saw a translucent black woman slap the candles off the mantle. It is uncertain why Sally is so connected with the back parlor. And what problem does she have with the candles? Could this be the room in which she was poisoned?

When Loyd Hall was first opened for tours in 1989 a murder mystery was performed to celebrate the occasion. The script was based on the murders and tragedies of the house and guests acted out the parts. The next day strange things started to happen at Loyd Hall. When staff tried to go to the third floor attic room where Harry Henry was killed they could not open the door, even though they had left it unlocked the night before. Somehow a large chest of drawers had been pushed up against the only door to the room from the inside as if by a supernatural force. Black birds were seen flitting here and there on the second floor of the house. Even stranger, the large grandfather clock, which had not worked in years, suddenly began to chime on every hour. The ghosts of Loyd Hall were not happy with the living making light of their tragedies. The murder mystery performance was never repeated.

While most of the unexplained activity occurs in the big house, overnight guests have reported strange occurrences in the guest cottages on the grounds. Keep the following list in mind when making reservations.

Restored 1800s Carriage House Suites ~ this former kitchen has two guest rooms, including the Magnolia Suite and the Camellia Suite. Here, in both rooms guests have reported smelling food being cooked at 4:00 a.m., a time when breakfast would have been cooked on a plantation. Could this be the spirit of one of the cooks of the plantation? Or is it the spirit of Sally Boston whose apparition is often associated with the aroma of food.

Restored 1800s Commissary ~ one woman reported hearing screams during the night while staying in this room, but this could have been the sound of coyotes near the property.

Minda's House ~ guests have commented on feeling their bed moving and have asked staff the next day if there were any earth tremors in the area. The answer is no.

The McCullough House ~ in this large suite a couple turned off all their lights and went to bed, only to wake up around two o'clock in the morning to find every light on in the cottage.

So, as Beulah Davis says, strange things do seem to happen at Loyd Hall Plantation. Staff members do not like to be in the house late at night and never want to be there alone. But really, they need not fear. Nothing bad has ever happened to anyone because of the ghosts; they just seem to want attention. Summer seems to be their most active time at Loyd Hall so keep that in mind when planning your visit.

Four lost souls inhabit Loyd Hall Plantation. Indeed, for many years the house seemed to drive away humans. Perhaps the unearthly residents explain the rapid turnover of this house during its middle years. After all it takes a special type of family to be able to share their home with a ghost, let alone four of them.

5

Chretien Point Plantation
A Gray Battalion of Lost Souls

A majestic plantation house stands in the small town of Sunset Louisiana; a house both proud and serene but also rife with intrigue, sorrow, and death. Lore and legend have swirled around Chretien Point Plantation for decades and the current managers and employees of the property find it challenging to separate fact from fiction. However, one fact about Chretien Point Plantation has been proven time and time again: there are ghosts here. A legion of spirits haunts the grounds and building, including a murdered pirate, a woman in white who comforts more than frightens, and the gray remnants of Confederate soldiers fallen in battle, still searching for a way home.

Joseph Chretien acquired the land on which Chretien Point Plantation was built in 1774 and 1776 through Spanish land grants. Joseph arrived from Quebec and his wife was from Nova Scotia. In 1818, Joseph's son and daughter-in-law, Hypolite Chretien and Felicite Neda started a cotton plantation on the land. Theirs was an unusual union as Hypolite paid $6000 for her hand. The agreement stipulated that should the marriage fail, she would keep the money. But the marriage did not fail and in 1831 construction of the grand house began. It is fabricated of brick made on-site and cypress taken from the family's swamp. The architectural style of the house is mixed. The prominent hipped roof and large galleries are characteristic of the French Colonial style, the columns are Tuscan, typical of the Greek Revival style of architecture, and the Adam fanlight windows are usually found in Georgian houses. An architectural oddity is that the river facade of the house is quite plain (this was the front of the house when the Chretien's lived there) while the back of the house (today used as the entrance) is much more elaborate. Presumably the Chretien's wanted their house to look less ornate than it actually was from the bayou, perhaps as a deterrent to would be river pirates.

The ground floor of the house has a formal dining room, a breakfast room (originally used as a food warming room and possibly as a children's dining room), a modern kitchen, and two guest rooms. The guest room, called the Wine Room, originally had a dirt floor to keep it cool and still has the original wine rack, which held four hundred and sixty bottles of wine.

The guest room called the Plantation Office has been used as a plantation office, a library, and a tack room for the horses of the plantation at different times. The original foyer has been cut in size to make room for two bathrooms for overnight guests. In the foyer stands the original ramp-knee staircase constructed of black walnut. It is believed that it was the model for the Tara Plantation staircase in *Gone with the Wind.*

The second story of the house is more elaborate than the first floor. This is where the family would have spent most of their time. The formal parlor is on this story along with three guest bedrooms. Historically these would have been the family's bedrooms. At the top of the stairs is the Magnolia Room and through the parlor are the East Room and the Prime Minister's Suite. In keeping with the French Colonial style there are no interior hallways in the house, but rather rooms open directly into other rooms. This method of construction helped with air circulation during the hot and humid Louisiana summers.

The current owners are only the fourth family to own the plantation. After the third generation of the Chretien family could no longer pay their taxes, the house was sold in 1899 to Celeste Gardiner Chretien's brother, Dr. Charles Gardiner. The plantation stayed in the Gardiner family in varying conditions, until it was purchased by the Cornay family in 1975. The Cornay family carried out extensive repairs and renovations to the house, and first officially opened the house to visitors. The current owners, Kristine and Kelly Nations, have done their own restoration work, and the house remains open for daily tours and as a bed and breakfast.

Ghostly activity seems to increase where tragic events have occurred. Violent deaths, children dying, or people passing with unfinished business seem to provoke hauntings. Strong emotions, both good and bad, can tie a soul permanently to a former home. The turbulent history at Chretien Point appears to be the catalyst behind the horde of ghosts who refuse to leave. In the following pages we will meet the ghost of a pirate, a specter woman in white, confederate soldiers trapped forever where they fell, a lady in black, and another woman in white who haunts a bridge. Strange occurrences are common at Chretien Point for both staff and visitors. It seems to be a focus of paranormal activity.

Many of the legends about the house and its occupants spring from the early days on the plantation. Disaster struck the plantation house during the yellow fever epidemic of 1838. First, Felicite and Hypolite's nine year old son Joseph died in 1838, and then in 1839, Hypolite himself died. From this day forward Felicite would always dress in mourning.

Felecite, was left to manage the plantation after her husband's death. Unlike most women, she did not use an overseer but ran the plantation herself. She has been described as a very independent woman, who liked to smoke, wear pants, travel alone, and enjoyed playing poker. She also is believed to have kept her husband's storied ties to the pirate Jean Lafitte. It is said that Lafitte and his henchmen used the grounds of Chretien Point for their illegal activities, including sales of smuggled and stolen goods on the plantation grounds.

After the death of Hypolite, Felicite became a target for thieves. He had left her three thousand acres of land and total assets of a value near $250,000. It is not surprising then that in the early 1840s Felicite was startled awake in the night by the hooves of horses thundering on her front lawn. Having few options, she decided to give up a jeweled necklace to appease the thieves. As she started down the staircase a man burst through the door. She displayed the necklace and the ruffian raced up to claim his treasure. As he neared, Felicite shot him square between the eyes with a pistol she had concealed in her dressing gown. He fell dead on the eleventh step. Hearing the gun fire, the man's accomplices fled away into the dark and the slaves rushed to the big house. The pirate's body was pulled down the stairs and stuffed into a small closet underneath the stairs. Legend goes that the house servants scrubbed the stairs and closet but could never remove the bloodstains.

Today there is carpet on the stairs and no blood visible around the edges, although previous owners swore to its presence. I did examine the closet under the stairs and while there are faint traces of stains, there are no longer dark bloodstains visible; time has finally taken its toll here. The ghost of this slain would-be robber makes his presence known in the house and is called Robert. While the previous owners, the Cornays, had many experiences with Robert, this particular spirit does not reveal himself as often these days.

Resting place of the murdered pirate

One of the most active spirits in the house is of a woman; it is not Felecite, but rather her daughter-in-law Celestine. It is perhaps fitting that Celestine should be haunting Chretien Point. Her portrait, which graces the space above the mantel in the parlor, has remained in its original spot since it was placed there in 1846. Celestine seems to like the comforts of the house, and seems to have fond memories of family meals as she usually appears in the dining room.

Margaret Belis is a tour guide at Chretien Point and witnessed Celestine in the dining room. On this occasion, Margaret was staying in the house overnight. She had locked the gate to the property and the house when another staff member, Nell Nations, told her there were teenagers roughhousing on the property and to stay inside. As Margaret walked toward the dining room she saw a person pass the window. She was going to open the door to let Nell know where the teenagers were when suddenly a figure appeared at the far opposite corner of the dining room. Her description of the encounter follows: "I was looking out that window and when I did there was a figure that moved from that corner of the room all the way across the furniture to me. I couldn't see her feet move; she glided kind of, all the way across the room through the furniture to right here [the doorway to the breakfast room]. She was wearing a long white dress, and I could see her little pink face and her hair in a bun. She was a young lady, well young to me, thirties or forties. It took her a few seconds to come across and I just stood here and looked at her. And I could see her until she got up to me, and when she got up to me, I couldn't see her anymore, but I could hear her, and I could feel her. She touched me just like this [lightly but firmly on the upper arms] and she said 'Don't go outside, don't go outside.' And so I stopped wanting to go outside, and as soon as I did, she disappeared, she went away. I think she was protecting me."

Margaret found the encounter with the lady in white to be comforting. Immediately she knew it was Celestine and felt protected. A contractor working on the property in March 2003 did not have the same reaction to his meeting with the woman in white. This man was working on the floors in the dining room and saw a "ghost lady" move diagonally across the room. Thinking he must be imagining this apparition, he tried to forget about it and went for lunch. After lunch the "ghost lady" appeared again gliding diagonally across the room. This was too much for this poor man and he packed up his tools and told the foreman he would not be coming back to work here as he had seen the "ghost lady" twice. We can imagine the woman in white was simply warning him about some danger in his work or keeping him away from a secret underneath the floors. Or maybe she was trying to prove she was there. Not only did he not believe in ghosts, but prior to the "visit" by Celestine he made fun of the staff, saying they were "silly women" to believe in ghosts.

In addition to revenants who continue to haunt the property, the house still bears scars of its violent past. During the Civil War, the Battle of Little Carrion Crow Bayou (also referred to as the Battle of Buzzard's Prairie) was fought on Chretien Point's grounds in October of 1863. The house was caught in the crossfire. The west column and the roof were damaged by cannon fire and the house riddled with bullet holes. One of these can still be seen in the breakfast room door. It was also during this battle that many outbuildings were destroyed. How the big house survived remains a mystery. Legend says that Hypolite II, sick and partially paralyzed from an ailment, was a Mason, and by sending the Masonic distress signal to Union General Godrey Weitzel the house was protected. Another legend states that his wife Celestine fed the starving Union soldiers with the family's chickens and hogs in return for leaving the home unscathed. Either way the house was preserved; indeed it was eventually commandeered by the Union forces and even used as a hospital.

The Confederate dead were buried in a large mass unmarked grave about one mile from the house. Confederate losses in this battle were shocking; many brave soldiers died, often suffering horrendous wounds inflicted from cannon and shot. Some of these tragic soldiers' spirits have no peace. Traumatized by their war deaths, this gray battalion of lost souls still patrols the property searching for solace, or escape, or their loved ones. Frequently, they are observed passing listlessly through the house and about the grounds.

Margaret Belis has caught glimpses of men in brown and gray fuzzy trousers and boots marching through the dining room. She has seen the men only from the back and waist down as they have no visible upper bodies, as if the soldiers march in both the spirit and earthly worlds. One morning around 4:30 a.m., Nell Nations was outside watering plants. She had the feeling that someone was looking at her, but saw nothing until she happened to look up at a tree near the swimming pool. Balanced on a long branch of the tree was a Confederate soldier with his left leg torn off from the knee down. Another time, Nell's nephew was staying in the Plantation Office and heard the creak of the armoire opening. Jerking around he was astonished to see an opaque confederate soldier with a big beard and a sword. He described this apparition as "The General" and was terrified by the experience. Once a family's youngest child refused to enter the Magnolia Room. She said there was a man dressed in some sort of military uniform with a belt, sword, and a gun. No one else in the family saw this figure, but the child was terribly frightened. After much coaxing the parents were able to get the child to stay in the room. Could this be "The General" that was seen in the plantation office? Other stories include seeing a grayish-white figure walking down the hill toward the pond, and the sounds of boots have been heard by the pond when no one is there. It seems the Confederate dead from this battle are unable to rest and are doomed to wander this battle site for eternity.

Odd and unexplained things happen everyday at Chretien Point Plantation and there are stories about hauntings in every room of the house. Objects often go missing only to turn up, sometimes months later. However, keys that go missing in the house never reappear. During a tour for about twenty teenagers Margaret reports that the lights in the parlor kept going on and off as if by unseen fingers. Margaret also notes that for a period of two to three weeks in October 2003 pennies appeared all over the upstairs. Oddly, she found only one penny in a room at a time, but pennies turned up as often as three times a day in different rooms. She discovered them on top of the pillow in the bed in the East Room, on the outside upstairs gallery, on the counter in the bathroom, and on the floor in the parlor. The pennies showed up whether there were guests in the house or not. And then suddenly this phenomenon stopped.

The dining room is one of the rooms in the house where ghostly noises are regularly heard. Often the staff in the house will hear footsteps in the dining room, doors opening and closing, and conversations in quiet voices. Upon further inspection the room is always empty. Sometimes footsteps are heard going up the stairs, followed by pacing in the parlor. Could this be Celestine? Or is it perhaps Robert, or maybe even Felicite? It is impossible to be certain.

We do not know which spirit is responsible for calling out the names of staff members from various parts of the house. Nell and Margaret report they will often think the other is calling them only to find out they have not. Nell has also heard her name called out on the grounds of the property when there is no one around. Theresa White, an overnight host at the plantation, returned to the house one night after locking the front gate. She was greeted by the question, "Theresa, are you back, is that you?" She replied "yah," but the person she thought had spoken to her was sound asleep. There was no visible source of the voice. This experience did not frighten Theresa; at least somebody in the house was concerned about her returning safely.

More than the employees experience strange events at Chretien Point. Guests slip on the eleventh step where the pirate was shot. Visitors report being touched by cold hands in the parlor. There is the smell of smoke in the fireplaces, which have been blocked up for years. All over the house, voices are heard, boots echo, and doors open and close when no one is there. The white sofa in the parlor always looks as if it has been slept on each morning, perhaps by Celestine whose painting overlooks it. An older woman with her hair in a bun has been seen on the upper gallery of the house, and also looking out of the parlor windows. While this woman initially appears human and has been mistaken for Nell's mother, she only appears when the house is completely empty. Doorknobs have fallen off by themselves, bricks on the walkway were moved and rearranged by unseen hands, and the sounds of horses have been heard in the driveway. Alarms go off in the house and in cars for unknown reasons, and visitors who arrive by motorcycle often are unable to restart their vehicle when it is time to leave. The mechanical problems experienced in the house and with vehicles have been attributed in the past to the revenge of Robert, the slain thief.

Each room seems to have particular events that happen. Keep the following list in mind when you make your reservations. You may have the experience of a lifetime.

Magnolia Room ~ this room, which was once the boys room, witnessed the deaths of several children over the years. Apparently the spirits of these children have lingered in their home. Quite often visitors who stay in the Magnolia Room say they feel someone at the foot of the bed touching their feet. Other times they experience jumping up and down on the mattress, like a child playing. Guests also report something pushing up against the bed or sitting on it. Sometimes they become aware of a child snuggling up next to them in bed, but there is never anyone really there.

This is the room where "The General" was seen and where a woman in mourning stares forlornly out a window, perhaps looking at the Bayou for a loved one who never returned. Tesa Laviolette, a staff member at Chretien Point Plantation, saw the lady in mourning, or My Lady as she calls her, four times in the window of the Magnolia Room. On three of the occasions Tesa was facing the pool when she shivered as if someone was watching her. Looking up at the Magnolia Room she saw a shadowy woman holding back the lace curtain. The woman was wearing a hat with a flat top and brim. A net came from the brim of the hat and formed a bow. The woman also had on a short black cape. Then the curtain dropped back and My Lady vanished; there was no one in the house at the time. The last sighting occurred when Tesa was going to the rear of the house and the woman in mourning was darkly outlined in the window. The curtain dropped and she hasn't been seen since. Tesa doesn't believe this was Felicite or Celestine, but that it was an older lady, somehow connected with the house. Make sure to look up at the Magnolia Room window when you are outside to see if My Lady is looking out over Bayou Bourbeux.

Prime Minister's Suite ~ at one time the nursery opened up directly into this bedroom. Guests report the sounds of little feet scampering across the room, the scent of cigar smoke, and a terrible moaning as if someone is in great pain or in deep grief. Perhaps these sounds are from Hypolite II who died in this room.

East Room ~ in this room guests have reported hearing sounds of moaning or someone gasping for breath, only to find their partner is sound asleep and no one visible is present. I had my own experience while staying in the East Room. I was resting on my side, not yet asleep, when I heard the sound of something breathing beside me. Knowing I was the only one staying on the upper floor of the house I did not turn and look. Then suddenly a very cold chill was in the air and I started shivering and shaking. I heard a woman's voice sharply call out, "Randall." Once the name was spoken the feeling passed, the temperature returned to normal, and there was no sign of anyone present in the room. The following day I asked who Randall was and no one in the house had ever heard of anyone connected to the property with that name. While nightly ghost visits are always suspect (how do we know the person wasn't dreaming or experiencing sleep paralysis?) I was definitely awake when this event occurred. It left me shaken and frightened until dawn.

Plantation Office ~ "The General" has appeared in this room. While staying here, a guest heard someone rushing up the stairwell followed by a bloodcurdling scream. Thinking something upset his sister staying upstairs, the guest ran to her, only to discover her sound asleep. Maybe this was a visitation by the pirate Robert.

Wine Room ~ the Wine Room is usually occupied by

the overnight staff member, but occasionally guests stay in the room. This room seems to evoke peculiar problems with clocks. Theresa White, who is an overnight host at the plantation, has had strange experiences in the Wine Room. While reading a newspaper in the room, a clock, which was stored in the armoire, startled her by ringing. Two things make this occurrence odd. First, when Theresa removed the clock and turned it off she realized it wasn't even ringing at the time it was set for. It was set for six-thirty, and it was eight minutes to six. Secondly, she had stayed in the Wine Room for the previous two nights and the clock had never gone off. As soon as Theresa put the clock back in the armoire, she heard someone come in the back door, her guest for the night. The guest had come to the front door, but no one heard her knocking so she went around to the back. Theresa assumes the clock rang to let her know the visitor had arrived.

A guest staying in the Wine Room also had a strange incident with her timepiece. She always carries a travel clock and had set it up on her bedside table before retiring for the night. During the night this guest attempted to check the time on her clock only to discover that it was not there, nor anywhere around the table or the bed. Later she found the clock, folded closed, under her pillow. So be warned if you end up in the Wine Room, you might just get an unexpected wake up call.

As an interesting aside, taking a short walk or drive from Chretien Point Plantation will bring you to a bridge that is reputed to be home to a spirit. Legend has it that Marland Bridge is haunted by a mysterious woman in white. Just who this woman may be is much debated. She does not appear to be Celestine, the woman in white of Chretien Point Plantation. In her book *Ghosts Along the Bayou,* Christine Word discusses the theories that it is either the ghost of a teenage girl whose murdered body was found near the bridge in the early 1970s, the ghost of a ten year old girl struck by a car on the bridge in 1922, or a ghost that is linked to the Civil War. The third possibility does have an association to the history of the bridge. Marland Bridge received its name as a result of the bravery of William Marland. He individually secured the bridge, which was being held by the 11th Texas Infantry, by charging with his team of eight horses and large artillery piece. This bravery earned William a Congressional Medal of Honor and the bridge has been called Marland Bridge ever since. At Chretien Point Plantation they also believe the ghost is from the time of the Civil War, likely a relative of Felicite, who was staying at the house. It is thought that her husband was killed during the Battle of Little Carrion Crow Bayou and in her distress she ran towards the bridge. She either inadvertently fell in the water, or threw herself in the water as a result of her grief. Now her spirit remains trapped at the bridge.

Regardless of who the woman in white is, the stories of her haunting remain consistent. Dogs and horses are said to be afraid to cross the bridge and many people who ride horses in the area must dismount in order to coax their horses across. Others have seen her figure gliding strangely along the bridge, sliding up to cars, and drifting along the bayou. The bridge was wooden until 1997 when it was reconstructed in concrete and metal. This change has apparently not stopped the visitations by this persistent spirit. During the work on the bridge her apparition was seen early in the morning by several of the construction crew, and they also heard strange sounds and had their equipment moved around. People still stop at Chretien Point Plantation regularly to report sightings of the woman in white. I was not so lucky as to catch a glimpse of her during my trip, but you may be.

Chretien Point Plantation is open for daily tours, as a bed and breakfast, for ghost dinner tours, and for special events. Ninety percent of guests report strange occurrences during their stay, the majority of whom have never heard stories of the hauntings. Chances are if you are brave enough to spend the night something will happen to you too. But don't be afraid. The throngs of ghosts at Chretien Point Plantation have never caused any harm and only seem to want attention. They have never scared guests enough to make them leave the house during the night. And besides, you do not want to miss the delicious full plantation breakfast in the morning.

Oak Alley Plantation

6

Oak Alley Plantation
The Lady in Black

T here is no doubt that Oak Alley Plantation is haunted; too many staff and visitors report sightings and unexplained experiences inside and outside the big house. Over the years, many people have talked about a mysterious lady in black, who sadly wanders the plantation. But there are other spirits who haunt Oak Alley. A young girl appears in the attic and second story of the house and the original owner peers intently into his former home, as if he is exiled outside eternally.

Before there was the plantation, there were the trees.

In the early 1700s a French or Spanish settler planted an alley of live oaks. The alley consists of two rows of fourteen oaks planted eighty feet apart, resulting in an avenue of a quarter mile leading to the Mississippi River. What happened to this pioneer, and why no house of equal magnitude was built to match the oaks, leading nowhere, has been lost to time. We do know that this original settler built a small cottage in the early 1700s where the mansion now stands. When the Capuchin Fathers arrived in Louisiana in 1722 to settle St. James Parish they commented on the oak trees, but made no mention of a home. Likewise when construction began on the current home in 1837 there was no mention of a previous residence on the property. The home of this early pioneer faded, but his true legacy, the oak alley, flourished. The alley of live oaks became a landmark along the Mississippi and even after Jacques Telesphore Roman finished his great home in 1839, the glorious tunnel of dark-green oaks dominated the property.

Francis Gabriel "Valcour" Aime, brother-in-law to Jacques Roman, and the king of the sugar planters in the South, acquired the land at Oak Alley in 1820. Valcour Aime and Jacques Roman exchanged properties in May of 1836. Roman's Classical Revival house was built with twenty-eight colossal Tuscan columns supporting a double gallery surrounding the home, mirroring the number of live oaks in the magnificent alley. Jacques' wife Celina called the plantation Bon Séjour, but the steamboat captains insisted on calling the plantation Oak Alley, and that is the name that has stood the test of time.

The two and a half story house has a simple central hall plan, with a hipped roof and dormers. Walls throughout the home are sixteen inches thick and the ceilings are twelve feet high. On the ground floor of the house visitors find three rooms and the elaborate entrance hall. The formal dining room with its mahogany dining table and shoo fly fan are on display along with gleaming china, silver, and crystal. The parlor displays the portraits of Celina and Jacques Roman, along with a piano and Empire furniture. These rooms have recently been restored and furnished according to an original household inventory. The third room originally contained a spiral stairwell, which was damaged beyond repair during the plantation's period of abandonment. Later the room was transformed into either a guest room or day bedroom. Its current use is to interpret the life of an overseer of a plantation and portions of the walls have been left exposed to show their construction technique.

Jacques operated a successful sugar cane plantation on twelve hundred acres of land, including twenty-four slave cabins, a hospital, an overseer's house, stables, a sugarhouse, and a sawmill. However, he did not enjoy a long life at his plantation, dying of tuberculosis in 1848 at only forty-eight years old, a mere nine years after the completion of his home. During their life together Celina had often fought with Jacques. She preferred the excitement of life in New Orleans, and he, life at Oak Alley. When he died, she grieved dearly and wandered the lonely plantation in veil, hat, long dress, and parasol of mourning black, as was the custom at the time. Celina was not interested in running the plantation and left the day-to-day affairs to an overseer. She spent money lavishly and traveled often to New Orleans, perhaps to escape the sad memories at Oak Alley. She deeply regretted fighting with her beloved husband and never really recovered from his untimely death.

Besides the early death of Jacques another tragedy would befall this first family of Oak Alley. The Roman's daughter Louise had her leg amputated. The story goes that her suitor came to visit her in a state of extreme intoxication. It would have been a social disaster for Louise to receive him in his inebriated condition. Fleeing away from him up the spiral staircase in the house, she tripped, fell, and the boning in her hoop skirt punctured her leg. The wound was severe and life threatening. Gangrene set in and an amputation was necessary. This infirmity led Louise to spend the rest of her life out of the social scene of a typical planter's daughter. She ultimately founded a Carmelite Convent in New Orleans, dying of natural causes in her eighties. In a macabre twist she was buried along with her amputated leg in the family tomb in New Orleans. How this leg came to be preserved has been lost, but presumably it was placed in the family tomb after amputation. In New Orleans family members are buried together and eventually their remains become as one in the tomb.

The Roman's only surviving son Henri took charge of the family affairs in 1859, but his mother Celina's never-ending spending and the lack of free labor after the Civil War forced the family to sell the plantation for a mere $32,800 in 1866. For the next sixty years, the house had a series of owners, experiencing both good times and bad. It is believed that the main house was even used as a barn for a period of time, and that its four-legged tenants damaged the original staircase beyond repair. From 1912-1917 the house was completely abandoned. The Hardins assumed ownership of the house in 1917. They began the important preservation and nourishment of the live oaks. Mr. Hardin also saved the alley from a levee setback, preserving the grandest of entrances. By 1924 the Hardins had experienced a series of misfortunes that required them to sell the property prior to the restoration of the big house. The house was deteriorating badly when Mr. and Mrs. Andrew Stewart purchased it in 1925. They renovated Oak Alley Plantation to its former grandeur. Mrs. Stewart lived there until her death in 1972 when the house and twenty-five acres passed to a nonprofit foundation. Seventy-five acres remain as a residential area around the foundation property, six hundred acres are leased for sugar cane cultivation, and the remaining four hundred and seventy-five acres continue to be virgin woodlands. The home is open for tours and there are guest cottages for those who wish to spend the night on the plantation grounds. Oak Alley Plantation has been designated a National Historic Landmark, the highest designation that can be awarded to a historic property.

Though the house has had many owners, a few residents have decided to make their stay at Oak Alley permanent. Today the ghostly lady in black roams the house and grounds of Oak Alley. During a visit to the plantation I interviewed two staff members about their personal experiences with the ghosts.

Sandra Schexnayder is the manager of the big house at Oak Alley. While she had heard many stories about the ghosts at the house it was only after three or four years working there that she personally experienced anything. Her first encounter was around 5:45 p.m. as she was waiting for the last tour to leave the house. The tour was in the upper gallery and Ms. Schexnayer had secured all the exterior doors of the house. Resting on the piano bench in the parlor after the long day, she suddenly noticed a woman dressed completely in black, including a black veil, walking toward the bottom of the staircase. She was confused and wondered how this woman could possibly have gotten into the house. As Sandra stood up, the figure turned toward her and abruptly disappeared. The woman was completely gone. Sandra was so frightened and bewildered by this encounter she had chest pains. She tried to think of a rational explanation for what occurred but could find none. This would not be her only experience with the famous lady in black. She continues to see glimpses of this mysterious ghost's skirt vanishing around a corner or floating up the stairs, and one night when she was alone in the house, cleaning a stain on the stairs, she felt an odd sensation at her neck. Turning, she faced the lady in black right behind her. These experiences have been very frightening, but overall there is not a feeling of dread or evil in the house.

Jane Landry, a tour guide at Oak Alley Plantation, also had a close encounter with the mysterious lady in black. In late 2001 she was sitting in the small kitchen area in the big house that guides use as a lunch and break room; it is attached to a small office and not seen on the tour. She was enjoying a cup of steaming coffee, all alone and content. Suddenly she felt there was someone behind her. This is a feeling that haunts Oak Alley. Ms. Landry looked back and inwardly gasped. In the doorway between the kitchen and office was the lady in black, standing perfectly still. This was a very short distance from her, no more than a yard and a half; she could nearly touch her. The lady was turned away from Jane, but was staring back over her shoulder. At the time all the hairs on Jane's arms were standing on end and she could not believe what she was seeing. Gathering up what courage she had, Jane swiveled her stool around towards the lady who did not move but just stared intently. In fear, Jane ran from the room and was astonished to find the coffee cup still in her hand.

Jane's description of the lady in black is the same as Sandra's. The woman was dressed all in black, with a black veil that comes half way down her chest, hands clasped, and hair up. From her close vantage point Jane noted that the lady in black is not wispy in form but seems solid with a distinct neck and facial features. It is not known for certain who this lady is, but based on the shape of the face Ms. Landry believes the lady in black is Louise Roman still lamenting her lost opportunities in life. Others think the lady in black might be Celina Roman forever mourning the loss of her husband. Whoever she is, the lady in black is the spirit who roams, the spirit who carries an aura of sadness, and who causes a sense of unease as if you are not alone.

While the lady in black is the most commonly seen ghost at Oak Alley Plantation, she does not appear to be alone in her haunting. Stories abound of a young girl, dressed in a white nightgown with blue trim. She has shoulder-length auburn hair and appears to be about twelve years old. Jane Landry saw this girl only once on a beautiful sunny day in 1997. Jane was sitting in a rocking chair in the upstairs hall when something caused her to look up to the third floor. (The third floor is unfinished and not open for tours.) She saw the red-haired girl in a blue and white nightgown move across the top of the staircase until she was out of her line of vision. Sandra Schexnayder has never seen this resident ghost of Oak Alley, but confirmed that many guides and visitors have reported sightings in the house. The identity of this ghost is unknown. Psychics say a girl around twelve years old died in the house, but there is no historical information to confirm that belief.

The third ghostly apparition that haunts Oak Alley is none other than Jacques Roman, the original owner of the house. On a particularly gloomy winter afternoon Ms. Landry was waiting inside the house for any potential visitors. Every once in a while she would look out the window to see if any visitors were approaching the house. One time when she did this, Ms. Landry was startled by the face of Jacques Roman peering in at her from outside the house. He was dressed just like he is in his portrait, which hangs in the parlor. Shocked by what she saw, Ms. Landry had to look away and when she dared to glance again he was gone. She confirmed that other guides at Oak Alley have also seen Jacques Roman glaring at them from the outside of the house. One can imagine that he is wondering who these people are in his grand house.

Besides these apparitions, many unexplained occurrences have taken place at Oak Alley throughout the house and grounds, which could be the work of one of the ghosts. It is interesting to think about the way the dead may manifest themselves to the living. For sightings, they may come back with the same body they had in life or in an ethereal, spectral, or shadowy form. They may be "seen" through actions such as turning on a light or moving an object. And they manifest themselves to our other senses, with a haunting fragrance, a rush of cold air over our skin, or cries in the night. At Oak Alley, smells of tobacco come from the room interpreted as an overseer's office, a room that Mr. Stewart used as a small bedroom to rest during the day. A guide heard the sound of a horse-drawn carriage outside the house as if visitors were arriving. Guides hear their names being called inside the house, and occasionally on the grounds. It is softly spoken, like someone is calling from a distance, yet the source is never found. Lights in the house have strangely turned back on after they have been turned off and the house alarmed for the night. A lamp in the upstairs hall used to shine brightly at odd times, and once when Ms. Schexnayder checked, it was not plugged in. Also in the upstairs hallway there was a rocking chair that on occasion rocked gently by itself, as if there was someone sitting there, musing on old times. Both the chair and lamp have been moved into storage. Such unexplained events can be quite unsettling to both staff and visitors.

Many guides and some visitors are afraid of the second story of the house. On this floor there is the master bedroom, a nursery, a sick room, and a bedroom left exactly the way Josephine Stewart decorated it, called the Lavender Room. The sick room displays medical supplies and is appointed as if there were an actual illness in the family. Consequently, mourning clothes and religious artifacts are on display. The nursery is decorated as a typical children's room with two small beds, a cradle, dressers, and dolls. Some guides have felt intense burning sensations and icy cold sensations in the sick room and the nursery. One guide felt a strong pressure on her chest and could not breathe as she led her tour group through the sick room. The same unnatural feeling overcame her in the nursery and she was unable to complete her tour. Ms. Landry describes a feeling of sadness that can overcome you in the Lavender Room or the sick room.

Ms. Landry also recounted a very recent experience (October 2003) with a strong overwhelming fragrance of roses in the Lavender Room, which was Mrs. Stewart's bedroom after her husband's death. The scent was so intense it was as if the entire room was filled with roses. Of course there were no roses in the room. Roses were Mrs. Stewart's favorite flower; perhaps she too was paying a short visit from beyond.

One of the most dramatic experiences occurred when a guide toured thirty-five Gray Line Bus passengers through the dining room. A candle fell out of its candle holder. The guide replaced the candle and restarted the tour, but it rattled to the floor again. The candle seemed to move with a will of its own, or as if thrown by a supernatural hand. There was no other explanation for it. The fall should have broken the candle, but it was unmarked, like new. The guide took the message and hurriedly moved the group along to the next room. Sometimes ghosts prefer not to be disturbed.

You can experience the ghosts of Oak Alley by spending the night in one of the cottages on the property. Stroll the grounds at your leisure and dine in the shadows of the oaks. But beware, ghostly activity has been known to take place away from the big house; overnight guests have placed calls to the office staff in the middle of the night seeking an explanation of the figure wandering the grounds that just disappeared.

Be sensitive during your tour or overnight stay at Oak Alley to what you see, hear, or smell, it just might be a visitor from beyond. These spirits do not harm, they are simply restless, roaming the home they cherish and love, seeking eternal comfort at Oak Alley.

San Francisco Plantation

7

San Francisco Plantation
A Ghostly Tea Party

Along the River Road you will find an elegant and colorful plantation home. Unique among the remaining plantations in Louisiana, San Francisco looks like a lady dressed up to go to a gala ball. The vibrant colors, elaborate styling, and matchless interior make this plantation unforgettable, not only for the living, as previous occupants of the house find their former home hard to forget and hard to leave. Beyond the extravagant furnishings, elaborate moldings, and décor this plantation houses other reminders of its past, the ghosts of a man and two little girls.

Edmond Bozonier Marmillion bought the land on which San Francisco now stands in 1830 and began work on his plantation home in 1853, with construction completed in 1856. Tragically Edmond would die that same year, never to enjoy his home. His son Antoine Valsin Marmillion had been traveling abroad and was bringing home his new bride from Germany. At first he believed the house full of flowers was to welcome his return, only to learn that his father had died the previous day. As Valsin's older brother Pierre had died in 1852 of yellow fever and his younger brother Charles was only sixteen years old, the responsibility of San Francisco fell to him. He and his wife Louise von Seybold were forced to stay on at San Francisco. What had been expected to be a short visit turned into a permanent arrangement.

In many ways San Francisco's floor plan is typical of French Colonial plantations, even if its décor is not. The first floor is constructed entirely of brick while the second is brick between posts. The first floor of the house has a central formal dining room and the service rooms of the house, while the second story has the formal parlor and bedrooms. The double gallery at the front of the house is dominated by Corinthian columns separated by arches; it opens to an elaborate double flight of stairs leading to the second floor. From the billiard room, two matching staircases rise to the second floor; one is original and the other a reproduction.

On the exterior, a decorative balustrade highlights the top story with rows of seventy-six louvered windows to ventilate the attic. There are two dormer windows on each of the four sides of the hipped roof. To top off this wedding cake style house, there is a widow's walk with a delicate rail. Cisterns flank the plantation house, with onion-domed covers. These are not only beautiful but were functional, providing a system to flow water into the attic where it was transported throughout the house to various sinks by an elaborate system of internal pipes. Thus this house had interior running water in 1856. In total the house is eleven thousand square feet and contains seventeen rooms.

San Francisco's exceptional architecture is matched by the unusual and stunning use of color. The shutters and portions of the cisterns are blue. The body of the house is peach and the trim is white. The style of the house has been called Steamboat Gothic, because it resembled the beautiful steamboats that passed along the Mississippi River. The house's architecture captured the imagination of author Frances Parkinson Keyes and she wrote *Steamboat Gothic* about a fictional family living there.

While the shape and detailing of the house can be attributed to Edmond Marmillion, the exterior and interior decorations are the work of Valsin and Louise. After four years in the house, Louise was unhappy with the pale, delicate colors and furnishings. In an attempt to appease his wife and make her feel more at home, Valsin told Louise to redecorate the home as she wished. Louise's German influence can be found in the elaborate and rich colors used throughout the home. Rooms boast walls of purple, brown, yellow, blue, and green. There are five painted ceilings in the home depicting cherubs, flowers, birds, and garlands. There are also hand-painted doors and faux-marbled mantles. Other doors and trim are cypress faux-grained to replicate various types of wood including oak, zebra, and maple. Ninety percent of the painting in the house is original. Faux-bois and faux-marble techniques allowed a greater variety of looks throughout a home, and when the craftsmanship was well executed, were very difficult to distinguish from the original materials. The decorative painting techniques used throughout San Francisco are excellent and wonderfully preserved.

Louise's beautiful restoration of the house came at a tremendous cost. Prior to the renovation the plantation was known as the Marmillion Plantation. After the restoration, Valsin would comment that he was *sans fruscins*, a slang French saying that meant "without a penny in my pocket," a reference to the high cost of the home. The name evolved to St. Frusquin. A future owner changed the name in 1879 to the more appealing San Francisco.

By 1870 Valsin and his younger brother Charles were running the fifteen hundred acre plantation, having bought out the interest of the heirs of their brother Pierre. In 1871 Valsin died at the age of forty-four leaving his wife Louise and brother Charles to run the plantation. Charles had served as a Captain in the Confederate Army and fought in four major battles, including Gettysburg. He was captured twice and remained a POW after the second capture until his release in April 1865. He suffered leg injuries during the war and was never well afterwards. In

1875 he died of syphilis at the young age of thirty-five. Louise managed to continue operating the plantation on her own until 1879 when she sold it to Achille D. Bourgere and moved back to Germany with her three daughters.

The Bourgere family owned the home until 1904 when it was sold to the Ory family. In 1927 there was a great flood of the Mississippi River. The dramatic and rapid change of flow of the Mississippi River eroded its banks. In order to ensure the safety of the surrounding areas a series of levee setbacks were required, and the levee itself was raised in height. Several plantations were destroyed during these setbacks, but the San Francisco Plantation house was saved, although its beautiful formal gardens were not so lucky. Today the River Road runs adjacent to the front of the home, making it impossible to photograph the elegant entrance hall. In 1954 the Ory family leased the house to Mr. and Mrs. Clark Thompson who first opened the house to the public. The Energy Corporation of Louisiana (ECOL) purchased the plantation in 1973 as an oil refinery, and extensive renovations began on the house in 1975. Marathon Oil acquired ECOL and completed the authentic restoration in 1977 at a cost of over two million dollars, returning San Francisco to its pre-Civil War splendor. Marathon Oil continues to support the cost of operating this historic house, and the San Francisco Plantation Foundation keeps the house open to the public. San Francisco Plantation was saved while others were lost, surely making its former owners happy, and perhaps this is why some phantom inhabitants have stayed.

San Francisco Plantation has long held the reputation of being haunted. Local residents of Garyville have always felt a supernatural presence on the property. This sense that ghosts are around is experienced at many plantations. It is like a tingling at the back of your neck, a vague unease, perhaps a shiver from an unexpected cold breeze, or an unexplained translucent image disappearing just after you notice it. These grand old homes seem to be awash with invisible currents from beyond. And the visiting spirits do not seem to want revenge but perhaps are motivated by an emotional need, a thirsting to relive past events, to feel comfort in their old homes, to luxuriate again in the humid lushness of Louisiana.

At San Francisco Plantation the most active ghost appears to be Charles Marmillion. The International Society for Paranormal Research (ISPR) undertook a six-year study of the hauntings in and around New Orleans in the early 1990s. San Francisco plantation was one of the sites they investigated and they concluded that Charles was haunting the property. The ISPR described him as a man with a mustache, reddish-brown hair, and a long brown coat. They concluded he is active in the bedroom in which he died, the dining room, and on the first floor. The house manager at the time admitted after the conclusion of the investigation that she had seen Charles in the house.

A more recent paranormal investigation in 2003 concluded Charles was present in his bedroom, other spirits were present in the attic, and an unknown woman was felt on the reproduction staircase. No staff members had previously

86

encountered an adult female spirit in the house and her identity has not yet been revealed or confirmed by any other source.

Some guides at the plantation feel a presence in Charles' bedroom. One former guide even reported that she felt a slight tugging on her skirt while in this room. She turned but there was nothing. Perhaps the bachelor Charles was trying to get this young lady's attention. While I was visiting Charles' bedroom my guide had some difficulty in opening the door, something that does not usually happen in this room. Was Charles trying to keep us out? Or was this simply a case of a door sticking in an old house?

Charles' presence is also experienced in the butler's pantry and the original stairwell. Some staff members and visitors feel an oppressive sensation while climbing the stairs, as if their breath is being taken away from them. One visitor said there is an everlasting supernatural party in the room at the top of the stairs, the loggia. She explained that the spirits sap the energy of those ascending the stairs to continue their party. But can you believe this? The loggia was in general a private area for the family, a more casual space. The entrance foyer and two parlors could be converted into a t-shaped ballroom by opening the doorways. This seems to be the more likely location of a party. Moreover, Charles is the only entity who has been felt in the house by staff members.

Some visitors have refused to stay in the butler's pantry, as they feel engulfed in an oppressive, overwhelming sensation, as if they are being smothered upon entering the room, and all their energy is being taken from them. They have left and waited outside for the tour to complete that room. One guide feels Charles' presence in the butler's pantry and stairwell area and says she has caught glimpses of his shadow when she enters this space.

Charles is also thought to be responsible for opening and closing doors in the main house. Perhaps he has returned to inspect his glorious house and enjoy its splendor. On a regular basis the front and back doors will open in the main house as if by unseen hands. While I was visiting the plantation the back door did open on its own, but it was likely the result of a cross-breeze from the front door opening. Apparently at other times there is no likely source for the doors opening on their own.

The front two doors of the gift shop also open on their own. Several employees have witnessed this phenomenon. It generally seems to occur when a staff member is in an office (the office area is in the same building as the gift shop with the rooms being to the left of the public space). They will distinctly hear the front screen door and heavy front door open and close, although it is not usually accompanied by footsteps. Upon inspection no one is in the gift shop or visible around the building. General Manager Kim Fontenot confirms this occurrence. In order to show me that the front door makes a distinctive noise I was told to stand in the back office near the copy machine. The front door was opened and then closed like someone entering the shop. The door has a distinctive squeaking noise that could not be confused with another sound. It was clear and easy to hear from the office area.

Could Charles also be responsible for the odd heating and cooling problems in the main house? Inexplicably areas of the house will be very hot or cold. When the maintenance worker inspects the HVAC system he always comments that it is working just fine, and there is no mechanical reason why certain areas are hot or cold. It is said Charles still roams his former home because of his long illness and untimely death. It is not usual for those who die in peace to return and wander the earth.

Charles is not the only spirit who haunts San Francisco Plantation. The spirits of two young girls seem to be trapped forever on the grounds. One staff member, a lifetime resident of Garyville, recalls that about twenty years ago a woman drove by the plantation every morning around 4:00 a.m. on her way to work. This woman said she saw two young girls playing jump rope in the front lawn. At this time the house was open for tours to the public and no one was living in the mansion. Yet this woman reported seeing the girls for many years.

Another staff member of San Francisco Plantation has also seen two young girls around six to seven years old playing on the property. She hears them laughing and has spotted them behind the gift shop or playing tea under a tree on the plantation grounds. She describes the girls as wearing white dresses, laying out a blanket and a tea set. The girls seem happy and playful. While she is the only staff member who reports having seen the girls, one day a visitor saw a similar sight. This visitor came running from the parking lot to the gift shop and excitedly asked,

"Did you see the two little girls?" The staff member probed her and the woman described them in the same way, dressed in white and having tea on the lawn under a tree.

Who the girls are is less clear. Valsin and Louise had five daughters. One died as an infant, another died at the age of two after falling down the stairs. The three other girls survived and only one, Emma, married. When Louise sold the property in 1879, she returned to Germany with her three daughters. Why only two of the daughters haunt the property and which two they are remains a mystery.

San Francisco Plantation has a richness to it that is hard to match at other plantation homes. It is unique in its color and ornamentation. Today the home is a National Historic Landmark, the highest designation given to a historic property, and one that is fitting for this lady. The restoration of the house uncovered a gem of a home, but as can happen when restorations occur, it seems to have stirred up a ghost, whose life was taken from him too quickly and who forever roams the halls of San Francisco, and two girls who have returned to play at the home they once loved. Visit San Francisco Plantation and you will see a home unlike any other in the state of Louisiana. If you are lucky, you just might catch a sight of the elusive ghosts of the house. These spirits seem mostly to be unaware of the living. It is as if we have intruded into their space and are gifted with a brief view of the vast deepness of that other world.

8

Ormond Plantation
A Cursed House

Ormond Plantation is entangled in mystery and intrigue. This centuries old Louisiana plantation is home to many tales of ghosts and hauntings. One night an owner disappeared under strange circumstances, never to return. Another was shot dead and his body strung up on the plantation grounds. Legends abound of buried treasure and the home is said to be cursed.

Pierre Trepagnier built Ormond Plantation on a Spanish land grant around 1787, making it the oldest West Indies style plantation home and one of the oldest surviving plantations in Louisiana. From the front it looks like a typical two story

Creole home with Tuscan columns supporting the balcony and colonnettes supporting the hipped roof. However, the balcony is only in the front and does not encircle the home, as was common with Creole houses. Some architectural historians believe there was originally a double gallery on the rear of the home, but no physical evidence remains. Ormond's floor plan is particularly unusual for a Louisiana plantation of this period with a four room central block and hallways on either side of this cluster of rooms. These hallways contain the stairwells of the home. The ground floor of the house has not only a formal dining room, but also a formal entrance parlor and a music room separated by a large arched doorway. All three rooms have gleaming crystal chandeliers. Transoms over the doors of the house have an unusual sunbeam pattern.

Ormond Plantation

The curse of Ormond Plantation is thought to date from 1798 when one of Trepagnier's slaves was brutally disciplined for attempting an uprising. In revenge this slave cast a spell against Trepagnier and his home. Regardless of whether we accept this story of a curse, tragedy did strike in 1798. Pierre was enjoying a meal with his family when he was summoned to the front door of his house to meet a visitor. This gentleman, dressed in the clothing of a Spanish official, had a private conference with Pierre. Pierre left with this "official" in his carriage and never returned. Upon inquiry, no one from the Spanish government had been dispatched to the house and no trace of Pierre was ever found. Mrs. Trepagnier moved out of the home a year later, unable to stay because of the memories it held of her husband. Giving up hope of her husband's return she finally sold the property in 1805 to Colonel Richard Butler. It was Butler who gave the home the name "Ormond" after his family's ancestral home Ormond Castle in Carrick-on-Suir, Ireland.

In 1809 Butler sold one third of his plantation to Captain Samuel McCutchon. In 1819 Butler turned all of his holdings over to McCutchon, who had married Butler's sister. Butler moved to Bay St. Louis, possibly to escape the yellow fever epidemic that gripped Louisiana. If this was indeed the case it was unfortunate. For in Bay St. Louis, Butler, the second master of Ormond, met an unfortunate and early death at the age of forty-three from yellow fever, along with his wife. McCutchon lived on at Ormond.

Either during the ownership of Butler or McCutchon two garconnieres (the separate sleeping area for young men of the household) were built on the property, symmetrically flanking either side of the house. The exact date of construction is debated to be either 1811 or 1830; McCutchon seems a more likely candidate for the addition as he had nine children. Each building is two stories high and has two bedrooms on each story with fireplaces. These two wings each have their own separate hipped roof and were built at a different height from the main house. The roof from the garconnieres slope down to connect with the roofs on the original home; passageways between the wings and the main house allow for circulation of air and contain stairways to connect the different buildings.

After the Civil War the McCutchon's were no longer able to operate Ormond Plantation. The plantation was sold and passed through four owners in the last quarter of the nineteenth century. In 1898 Ormond Plantation was bought by State Senator Basile LaPlace. LaPlace's stay at the plantation was short, as he too met a violent, tragic death. LaPlace was an adversary of the Ku Klux Klan or the "White Caps." There is some confusion surrounding the "White Caps," in some locations they were a separate organization from the Klan while in others the Klan were called "White Caps" because of their peaked white hoods. Like Trepagnier so many years before him, LaPlace was called out of his home one night in 1899 as he was eating dinner. His bloody body was found later hanging from a large live oak tree on the front of the property, riddled with bullet holes. His killers were never apprehended.

In 1900 the plantation passed from LaPlace's widow to

the Schexnaydre family who lived there until 1926. Ormond had numerous tenants throughout the 1920s and the 1930s and began to show signs of decay, until Mr. and Mrs. Alfred Brown finally purchased it in 1943. The Browns restored the house, enclosed the stairways, and built two one story back wings from each garconniere. After Mrs. Brown died, Mr. Brown sold the property to a real-estate developer who planned on turning the plantation home into a clubhouse for a golf course to be built on the property. Luckily he stopped his alterations in 1971 and the land was sold in 1974 to Mrs. Betty LeBlanc who spent twelve years tying to return the house to its original state. She passed away before her restoration was completed and her son Ken Elliott took possession. He completed the restoration and opened the house to the public for the first time. Today new owners live in one of the garconnieres and keep the house open as a bed and breakfast and for tours.

Ormond is furnished with Spanish furniture dating to 1850 chosen by Betty LeBlanc. One of the chairs in the dining room is interesting and perhaps fitting for this house; called a castle chair, it has gargoyles on the armrests. The purpose of the gargoyles is to keep away evil spirits. This was a sensible precaution considering the turbulent past at the plantation.

With such a volatile history, it seems only natural that Ormond Plantation should be haunted. Indeed ghostly legends surrounded the plantation as early as the 1880s when a ghost story was published in The Times Democrat in New Orleans. The story tells of an eerie encounter by a man who was sent down to the river to check on an arriving steamboat. Before he reached the landing, the steamboat hurriedly sped away, disappearing in the gloomy night. The man wondered if something had been left behind. He searched in vain, but there was an uneasy feeling in the night. As he returned to the house under the shadowy moss of the trees, the feeling increased. Just then the man ran into something in the darkness, something darker than the night itself. He stopped short in panic. The blackness shrank to a small spot and blinked out. He ran frantically back to the plantation house, swearing he had just seen a ghost.

Later that same night another man was alone in the center block of the house, pondering over the strange occurrences of the steamboat and the ghost. As he sat with these thoughts, he heard a sound like the rustling of wind through the live oaks moving through the house. It seemed to swirl in the adjoining room and then a darkness like smoke squeezed through the door's keyhole into the room where he sat motionless. A dark figure formed and glided about the room behind him. Suddenly a cold fleshless hand touched his face and he cried out in horror.

Is Ormond Plantation still haunted to this day? During my tour of the plantation the guide talked about the mysteries and curse of Ormond. I asked about ghost stories and while she had none herself, she said other staff at the plantation believe the Doll Room to be haunted. This room is full of dolls collected by Betty LeBlanc and is painted pink. Staff have reported seeing a young woman looking out of the windows towards the road and the river. They say she may be one of Trepagnier's daughters, forever awaiting her father's return home, anguished in the afterlife as to his unknown fate. Another guide has reported seeing a man dressed in circa 1900s clothing wandering the grounds in back of the plantation. The identity of this man has not been confirmed, but it is easy to wonder if it could be one of the unfortunate masters of the plantation.

Trepagnier, Butler, and LaPlace are all thought to have hidden their money somewhere on Ormond's property. None of the treasures of these men has ever been found. Perhaps the gentleman who strolls the grounds is trying to show us the location of his buried treasure, or maybe he still guards the fortune from his grave.

Today, Ormond Plantation has been restored and you can stay in one of the three bed and breakfast rooms or take the tour of the old home. If you do, you can see this cursed house and the location of these mysteries first hand. As you leave take one last look, you just might see the ashen face of a young girl pressed against a window, despairing for her father's return. In a house with a past like Ormond's the spirits rarely rest in peace.

Destrehan Plantation

9

Destrehan Plantation
Brothers in League with a Pirate

In Destrehan, a short distance from New Orleans, you will find the oldest documented, intact plantation home in the lower Mississippi Valley. During its history this beautiful house suffered terrible damage from fortune hunters, who may have been allured by the ghost of Jean Lafitte and his long lost buried treasure. However, two other spectral figures are most commonly seen here. One is a tall, thin man with hair like gray mist and the second is another man, who wears a cape and is missing an arm.

Destrehan Plantation was built from 1787 to 1790 for

Antoine Robert Robin de Longy by a man named Charles, a free mulatto. The plan for the house's construction called for a home sixty feet long by thirty-five feet wide in the French Colonial style with six chambers on each of the ground and second floors. The double-pitched hipped roof has three small dormers on the front and one on the rear to help ventilate the house during the hot summers. Originally columns and colonnettes supported the galleries, which surrounded the home. Typical of Louisiana French Colonial homes there were no interior hallways and access to the upper floor was through a flight of stairs on a side gallery. The house was made of brick covered with plaster and painted to resemble stone.

In 1802 Jean Noël Destrehan purchased the home at auction for $21,750 (his wife, Claudine Eleonore Celeste, was a daughter of the original owner). Sometime after 1808 he added matching two story wings to either side of the house, which were used as garconnieres. Another staircase was also added at this time opposite the original staircase. The most significant improvement that Destrehan made was changing the crop from indigo to sugar, which resulted in the plantation becoming extremely profitable. Jean Noël died in 1823 and his wife died a year later.

In 1825 Destrehan Plantation was sold again at auction to Stephen Henderson, who was married to Jean Noël Destrehan's daughter Elenor Zelia, for the princely sum of $186,971. The Henderson's never lived in the plantation, preferring their elaborate home in New Orleans while an overseer managed the plantation operations. There were disquieting rumors that the couple's marriage was troubled. He was twice her age (she was only fifteen when they married) and they never had children. The marriage was likely arranged by her parents; Henderson's great personal wealth meant that Destrehan would be kept in the family. (Louisiana follows Napoleonic laws, and there is forced inheritance for all children. If parents have more than one child, fifty percent of their estate must be divided equally among their children. This often required selling homes in order for all the children to receive their portion of the inheritance. One heir would often buy out the portions of his siblings in order to keep the home.) In 1830 Zelia traveled to New York and expressed a wish to change her will, leaving her personal estate not to her husband but to her younger sister Louise Odile Destrehan. (Her husband had given her a third of his total assets valued at $242,000 when they married in 1816. This property was her personal estate.) Zelia died mysteriously from unknown causes in a New York hotel before being able to change her will officially. Her estate was left to Stephen.

When Stephen died in 1838, his will caused a family uproar and litigation ensued. Instead of leaving the property to the Destrehan family he decided to have the plantation become the center of a town where a factory would produce clothing for slaves. Stephen also stated that twenty-five years after he died, his slaves should have the option to return freely to the new country of Liberia in Africa. Given the average lifespan of a slave this was not terribly generous. Still he did stipulate how the slaves should be treated during the twenty-five years; all children born after his death were to be apprenticed to learn a trade. The courts did not overrule the twenty-five year stipulation, but ironically twenty-five years after Stephen's death, the Union Army had already freed the slaves. The courts agreed with his heirs and overturned the other portions of his will. Henderson's heirs sold the plantation in 1839 for $185,295 to Judge Pierre Rost, husband of Louise Odile Destrehan. In another twist, Louise would ultimately live in her former childhood home, fulfilling the dying wishes of her sister.

The Rosts decided the home, which had not been truly lived in since 1823, needed serious remodeling. The back gallery was enclosed to make a proper entrance with the two winding staircases now on the interior of the home creating a magnificent entry foyer. The thin colonnettes were replaced with brick Doric columns more in keeping with the Greek Revival style. Elegant French doors replaced the old windows and doors, and beautiful yellow or white marble fireplaces replaced earlier ones. The small dormer windows remained, but were fitted with shutters.

In 1861, Jefferson Davis, President of the Confederacy, appointed Judge Pierre Rost as Minister to France. The Judge brought his family along and they all escaped the hardships of the Civil War. His diplomatic post ended in 1862, but the family stayed on in France until 1866. While the family was gone the Federal Army seized their property, including Destrehan Plantation, and turned it into a Freedman's Bureau, paying the freed men and women to continue to operate the plantation and learn trades. The Union Army also opened a school and hospital on the grounds. When the family returned Rost was able to demonstrate he had not supported the Confederate cause since 1862 and was merely residing in his native France. The plantation was returned to the family, however a school on the property continued to serve freedmen. Pierre and Louise never lived again at Destrehan, residing in New Orleans for the remainder of their lives.

Judge Rost died in 1868 and when Louise died in 1877 the plantation was sold to their youngest son Emile. He grew sugar and rice successfully until 1910 when, having no surviving heirs, he sold the plantation for $95,000. The Destrehan Planting Company was formed, but was forced to sell the plantation in 1914 for only $43,000 when the sugarhouse burned. The declining value of the plantation is incredible. After 1915 the plantation land was turned into an oil refinery and operated under several companies for over forty years. In 1958 the refinery was dismantled and the property was abandoned. Hurricane Betsy caused further damage to this once proud home in 1965. After years of neglect the River Road Historical Society assumed control of the home in 1971 when the Amoco Oil Company donated it. The Society has devoted years to restoring the house and today they run the plantation, and perform an excellent job of interpreting the evolution of the house. One room has exposed walls to illustrate construction techniques and contains models to show the alterations over the years. The tour of the home also includes a short video describing the history, the evolution of the house, and talks about the terrible damage done to Destrehan by treasure seekers. The video alludes to the idea that perhaps the ghosts of Destrehan helped to make sure the house itself didn't burn to the ground.

One name associated with Destrehan Plantation is Jean Lafitte. We have already met this gentleman rogue at Chretien Point Plantation, and will see him again in upcoming chapters. The legends of this pirate's escapades and his fabled buried treasure are common threads in Louisiana's folklore and plantation history. Jean Lafitte sold stolen slaves along the Mississippi River to plantation owners. Surely Lafitte trafficked with the owners of Destrehan Plantation. The two most likely candidates for this are Stephen Henderson and his brother-in-law Nicholas Noël Destrehan (son of Jean Noël Destrehan). Legend has it the pirate was a frequent guest at the plantation when Stephen Henderson owned it. We know that Henderson never actually lived at the plantation, establishing a perfect alibi to cover secret meetings with the famed pirate.

Others link the pirate to Nicholas Destrehan, although little evidence exists and Nicholas had enough traumas in his life. His first wife died young, early in their marriage. Shortly after losing his beloved Victoria, he lost his right arm in a horrid accident. As a gallant, he wore a cape everywhere. One day it became entangled in a piece of sugar cane machinery, which mangled his hand, threatening his life. To save himself Nicholas cut off his right arm. He miraculously survived this brutal wound. In 1826 he married again and established his own plantation across the Mississippi from Destrehan on land previously owned by his grandfather Destrehan de Beaupre. He began construction on his own home, called the Destrehan Mansion (creating some confusion by having a property with the same name as his father's) but in 1836 misfortune struck again, as his second wife Louise died from yellow fever. The house was never completed. But his fields were well tended and a canal formed by his grandfather helped with irrigation. A legend says that Jean Lafitte and his followers used this canal to escape persecution and seek refuge with Nicholas Noël Destrehan at his Destrehan Plantation.

A confusing set of events and certainly there is little proof that Jean Lafitte spent time at either Destrehan Plantation with Stephen or Nicholas although there is often an element of truth in legends. As for buried treasure, although it is unpopular to say so, it is unlikely that there is any buried fortune belonging to Jean Lafitte. In his later years Lafitte was destitute and undoubtedly would have lived well if he had money. Nonetheless Lafitte's treasure has been linked with countless sites, Destrehan among them. Indeed, legends say his ghost has been seen pointing at the house in an attempt to indicate where he left his elusive loot.

When the River Road Historical Society took over the house in 1971 they declared it exorcised of ghosts. An article which appeared in the New Orleans Times-Picayune in 1994 stated that the Board of Directors of the River Road Historical Society's stand on ghosts was that they neither admitted nor denied their presence at the plantation. And who can blame them? The house was nearly destroyed when it was abandoned for thirteen years, as vandals attempted to find the treasure of Jean Lafitte. Complete interior walls inside the plantation were removed, and of course nothing was found.

Stating that there are no ghosts at a house and having it be free of ghosts are two different things. Once hauntings start at a plantation it is difficult to put them to rest. Apparently ghostly activity in the house reached a peak during the restoration in the 1980s. Such a phenomenon is not uncommon. Spirits seem to be awakened by major disturbances in the homes they once occupied and often choose to make their presence known whether they approve or disapprove of the work. One assumes the ghosts at Destrehan Plantation were active to express their pleasure at the extensive restoration work by the Society, which returned the house to its original splendor. Now it is an elegant lady of a house, formal and proud of her lineage, assured of her place in the royalty of Louisiana plantations.

While restoring Destrehan, painters often heard footsteps in an adjacent room only on inspection to find the room empty. Three employees were interviewed for an article by the Associated Press, which appeared in the Times Picayune-The States-Item on Nov. 1, 1980. All three had on multiple occasions seen the apparition of a tall thin man in dark clothes, who materialized and then dematerialized before their eyes. None of the women believed in ghosts prior to this encounter. They were shaken and frightened by the experience. The descriptions of the man were consistent and the encounters similar. The women generally saw the man at the end of the day, after the gates to the plantation were locked. The man was usually on the front porch, looking into the house from the outside. When the women opened the door to let the man in, he was nowhere to be seen. Other reports include unexplained cold zones in the house, lights blinking on and off as if by spectral hands, and discarnate voices echoing through empty halls. Occasionally photographs revealed figures not in the room at the time the photo was taken. One such photograph from 1984 used to be displayed in the plantation gift shop. The photograph is said to have exposed a disembodied head in a mist behind a tourist in the foyer area, reportedly one of the most active areas of the home.

Over the years there have been numerous accounts of the hauntings at Destrehan. Some accounts claim that Stephen Henderson haunts his former home, perhaps forever furious that his final wishes were ignored. Sometimes Jean Noël Destrehan has been seen joining tour groups. One woman even claims to have spoken to him after he asked her what she was doing in the home. She explained that she was curious if one of her ancestors had been a slave on the property. The well dressed man with gray hair and sideburns calmly informed her that no one by the name she was seeking had ever worked at the plantation. The trouble was this was not a guide in interpretative clothing, but is believed to have been the ghost of Jean Noël Destrehan himself.

One of the most commonly seen spirits is thought to be of Nicholas Noël Destrehan, who is described as wearing a waistcoat, a shirt with a high color, and a cape, missing his right arm. Visitors have reported seeing this gentleman wandering the enclosed entrance foyer, which was originally an open gallery, and on the front porch and gallery. As we know Nicholas owned his own plantation across the Mississippi, but perhaps, as his own plantation no longer exists, he returns to Destrehan, and continues to mourn the loss of his beloved first wife, next to whom he was buried in a grave that bordered the fields of Destrehan.

Is Destrehan Plantation haunted? The stories that have passed down through the years seem to suggest that something unusual is experienced at the home. During my tour of the house I questioned the guide about ghost stories at the plantation. She said she was not sensitive to such encounters and nothing had occurred to her during her employment. She did say that many visitors experience unexplained things and insist the home is haunted. She acknowledged that a common apparition is Nicholas Noël Destrehan on the gallery of the home. Does Jean Lafitte haunt Destrehan? If Jean Lafitte is anywhere it is likely his own former cottage, Lafitte's Blacksmith Shop in New Orleans, now an intimate bar with reported sightings of its own.

Destrehan Plantation is open for tours, which illuminate the history of the house. But if you are observant you just might catch a glimpse of a man in your tour group with a high collar and a cape, or a distinguished older man with gray hair, like mist off the bayou. If you do, don't be alarmed, although for many of us there is a kind of singular thrill in being a little afraid at the thought of ghosts. It is possible Jean or Nicholas Destrehan is enjoying a tour of their home and grateful that the River Road Historical Society took the pains to restore it. You can shudder a little and smile, knowing you too have seen the otherworldly inhabitants of this plantation home.

La Branche Plantation Dependency House

10

La Branche Plantation Dependency House
A Spectral Horse

Along the River Road in St. Rose, Louisiana is the La Branche Plantation Dependency House. On this property you will find an impressive, two hundred year old garconniere and a slave quarter. These are the only dependencies that remain of a once prosperous sugar plantation. The large plantation house itself burnt during the Civil War. Today an alley of ancient live oaks leads to the eerie foundations, reminding us of the grand past of this plantation. But a tour of La Branche will reveal there is more remaining than merely these outbuildings, foundations, and live oaks. Former residents and visitors have a hard time saying goodbye and are still visiting from beyond the grave.

Johann von Zweig immigrated to Louisiana from Germany in the early eighteenth century and settled on the East bank of the Mississippi in what was known as the German Coast, because of the large number of German immigrants. The Ursuline nuns selected Suzanne Marchland to be his wife. Since von Zweig was a resident in a French area with a French wife and his name was difficult to pronounce, it was changed. Zweig translates to "branch" so he became Jean Baptist La Branche. Jean and Suzanne had six children Genevieve, Michel, Marie-Louise, Jean, Suzanne, and Alexandre. The family became successful sugar planters. With the death of Suzanne in 1780 and Jean in 1781 the property was sold and divided among their descendents.

It is believed that the La Branche Plantation's large house was built in the early 1790s and the dependencies were built in 1792. Alexandre constructed the big plantation house, which he called Barbarra Plantation. (Since the plantation has been called La Branche for years, that name will be used in lieu of Barbarra.) Alexandre was a Justice of the Peace and a militia captain. He and his brother Jean created a successful partnership on their plantation from 1796-1808. Jean lived in one of the two garconnieres on the property. The 1804 census shows that Alexandre La Branche lived on the property with his wife, four sons, five daughters, three grandchildren, and his brother. Alexandre had seventy-one slaves and his brother had fifteen slaves. As the family grew, they acquired more plantations, leading to an impressive portfolio of land on both banks of the

Mississippi River prior to the Civil War.

As an interesting aside one of these was a large plantation in St. Charles Parish owned by Michel La Branche's son Jean-Baptist (1777-1837). He built the La Branche house in New Orleans in 1832 at 700 Royal Street. Today the building is under restoration, but was recently the Royal Café, and is believed to be haunted in its own right by the widow of Jean-Baptist and his mistress Melissa. Because of the beautiful ironwork on its galleries this is the most photographed building in the French Quarter.

It is believed that Alcée Louis, one of Alexandre La Branche's sons took over La Branche Plantation after Alexandre's death. He fought in the Civil War and died in Hot Springs, Virginia on August 17, 1861, perhaps at the Homestead Hotel, which was the site of a war time hospital. The cause of his death, whether from disease or wounds, is unknown. Louis had no children. During the Civil War the main plantation house burned to the ground. The current owners of the plantation believe the La Branche family burned their home rather than have it be taken by Federal troops. This is a possible scenario, although there was no fighting in the immediate area, but federal gunships traveled up and down the Mississippi burning plantation homes. Perhaps what really happened to the house will remain a mystery.

The size and craftsmanship of the remaining garconniere at La Branche Plantation is magnificent. The big house of the plantation must have been truly extraordinary. In *Louisiana*

Architecture, 1714-1830, Fred Daspit includes a drawing of the main house at the plantation, which is a typical French Colonial raised home. The home appears to have had massive Tuscan columns supporting the double galleries, which surrounded the home. Over the entrance door was a fanlight window. The roof was hipped with three front dormers. The style of the main house is reflected in the only remaining garconniere.

The garconniere is of solid brick construction, with four equal twenty-foot square rooms. Three sides of the dependency house have galleries with substantial Tuscan columns, likely a reflection of the architecture of the big house. The dependency is on the National Register of Historic Places because of its rarity as a plantation dependency and its exceptional Federal woodwork, including the four original mantelpieces. It is both unusual for dependency houses to survive on their own without the big house and for them to be as substantial in size and elaborate in detailing as La Branche.

The Lentini family of Kenner, Louisiana has restored the plantation dependency to its former glory. It is furnished with period furniture and open to the public for tours, as is the remaining slave quarter on the property. Prior to the Lentinis, the dependency was owned by seven different families and had numerous renters. The last of the seven, the Mattingly family from New Orleans, bought the property in the 1940s principally as a breeding ground for racehorses and renamed the property Idle Horse Farms. Over the years the Mattinglys spent less and less time at the plantation, and even though there was a caretaker, he was elderly and unable to tend to all the horses, buildings, and grounds. Ultimately the property fell into disrepair. The Lentini family bought La Branche Plantation from the Mattinglys in 1983.

Ghostly activity abounds at the La Branche Plantation Dependency. There are sightings of spirits, including a unique spectral horse, strange cries as if from a suffering baby, and evanescent smells of smoke and perfume as if an elegant gentleman and his lovely lady saunter invisibly past you. There also seem to be poltergeists as household objects move on their own, lights flash mysteriously, and trees billow smoke.

One curious story about La Branche Plantation concerns the Mattinglys. After the end of the Second World War, the U.S. Army claimed Nordlicht, Hitler's horse, as a spoil of war and transported the horse to the United States. In 1948 Dr. S. Walter Mattingly purchased and brought Nordlicht to La Branche where he spent the last twenty years of his life siring numerous offspring. Nordlicht was buried at La Branche Plantation and there is a marker to designate his grave. When a worker for the plantation was planting a crepe myrtle tree in this location in the mid 1980s, he was horrified to discover bones underground, not realizing that they were of the horse. Perhaps it would have been better not to disturb this grave as everyone who lives on the property has seen a ghost horse since the grave was disturbed.

The stories of the ghost horse's visitations vary, but they all end the same; the horse disappears instantly; there is no earthly explanation. Lisa Lentini, director of the site, explains one of her experiences with the ghost horse. "My sister has horses. And I was about to leave one day and there was a horse outside of the barn. And I was like, oh well I can't leave that horse standing there; it could go out into the road and get killed. I went back there, and I touched him on the nose, and said you need to get back in, I closed the fence, he ran to the barn, and he was gone. There were no horses out of the stalls, the horse just vanished." Other family members have also reported seeing a brown horse with a white nose vanish before their eyes both near the barn and near the front of the property close to the grave of Nordlicht.

A spectral horse is unusual indeed and perhaps for this horse there is no place in the equine heaven. But there are further apparitions seen at this plantation. Two wandering spirits have been spotted on the grounds and additional ghosts make their presence known through our other senses.

Salvador Lentini, a staunch preservationist and no-nonsense former Kenner police chief, does not believe in ghosts. However that does not stop him from telling the story of the time he saw a boy who disappeared oddly. He describes the encounter, "It was about seven thirty at night and I was coming from the back [of the property in my truck] and in the road there was a little boy [standing]. He had a turtleneck long sleeved sweater on. It was, you know, a sweater of the past. I said I will be with you in a minute." The boy seemed out of place, but did not appear to be in distress. But when Salvador got out of his truck to tend to the boy, he was gone. There was no trace of him. The area where the boy was sighted is very isolated. It is in back of the dependency, on a stretch of road near a barn, quite a distance from the main road. There was no place for a young boy to go on the property, and no place where he could have come from. Salvador is still not convinced that what he saw was a ghost; he believes that it could have been an optical illusion, but does admit that strange things seem to happen at La Branche Plantation.

Blossom Lentini, Salvador's wife, also saw a ghost on the back road of the plantation near the barns. In this area of the property there is a small building with two rooms, one of which is the public washroom. The other room holds some artifacts including a bathtub reputed to have belonged to Richard Taylor, son of President Zachary Taylor. The bathtub originally resided on the opposite bank of the Mississippi at Fashion Plantation, which is now destroyed. Strangely, there is also a tombstone in this building belonging to a Balthazar La Branche (1816-1853). It is not clear how he is related to the La Branche's of the plantation. Through my research I did find a B.D. La Branche listed on the 1850 census living on the plantation of the widow D. La Branche. He is not listed on the 1860 census. Balthazar was probably a nephew of Alexandre. The tombstone fell over at the old cemetery in Destrehan and because he was a La Branche the Lentinis were asked if they wanted it. The tombstone now stands unceremoniously behind the bathtub, a very odd couple, causing a very strange feeling in visitors.

"There is something strange about this road."

One night at seven thirty in late 2002 Blossom was walking along the road from the back of the grounds toward the front of the property. (She was in the same location as Salvador was when he saw the boy years earlier.) She describes the encounter as follows: "I saw a man coming out of that building...I thought maybe he was using the bathroom at first because we have bathrooms there, but he was coming out of that second door [the room containing the tombstone and bathtub] and I saw him as plain as day. He had on a mounted police hat, a hat with a peak on the top, he had a hat like that, and he had navy blue pants, and a navy blue shirt. I never will forget it. He walked across there, and when I got to the end, I walked all around and I didn't see a soul. There is something strange about this road here." What makes Blossom's encounter even more inexplicable is that not only did the man completely disappear, but also the door that he walked out of was locked, as this event occurred after the site was closed for the day.

There is one original slave quarter on the grounds, which the Mattinglys moved closer to the front of the property. The Lentini family and visitors often smell cooking from the slave quarter, as if hearty meals are still being prepared there. The tang of cigar smoke is in the air around the slave quarter, in back of the house, and all over the property. We can imagine former planters inspecting their domain invisibly, savoring an after dinner cigar.

Another unexplained sensory experience is the lush scent of perfume. This occurs nearly everywhere, inside and outside the dependency house, and all around the slave quarters. Lisa Lentini describes the scent as an old style perfume, one that is rich with flowers, as if a huge invisible magnolia is in bloom with a mass of fragrant white flowers. However there are never any flowers near the fragrance. Visitors have also commented and asked about this overpowering scent of perfume. The heady scent occurs instantaneously and then as suddenly disappears. Could this be the former ladies of the plantation coming back for a quick peek at their home, leaving a tantalizing reminder of their beauty and grace?

Many odd activities have occurred in the garconniere itself. Most of these strange events happened prior to opening the house to the public. The Lentini family bought the property in 1983, but did not open the site to the public until 1992 and did not move to the property until 1993. Prior to this, Blossom would come by and check on the site a few times a week. Every time she arrived she would find the mantle candlesticks on the floor. At first she thought they fell when the door was closed, but it happened every time and nothing else was ever on the floor. To further add to her confusion, several figurines would be turned around when she came to the house. She would set them in their proper places only to find them turned around again on the next inspection. One time a rocking chair was moved and positioned close by a bed, as if a spirit was comforting an ailing loved one in the bed. Blossom suffered through the lights going off in the attic as if controlled by unseen hands. Blossom found this deathly frightening. The attic seems particularly hostile as suggested by the time Blossom discovered the attic door wide open; it is always locked. She climbed up to see if anything was wrong, and when she got to the top of the stairs she fell down breaking her pelvic bone. The strange thing is she felt she was pushed away from the door down the stairs. Perhaps one of the attic ghosts who haunts the garconniere does not wish to be disturbed and sometimes resorts to violent warnings. Interestingly, since the house has opened for tours the poltergeist-like activity has lessened, and only faint voices, cries, and spectral smells are occasionally experienced in the garconniere.

One of the rooms of the garconniere is haunted by the

anguished cries of a baby. This may be explained by the story of a woman, who rented a room in the garconniere in the late 1920s. She said that she had twins in this room and one of them died as an infant. Could the cries be from this lost baby who still needs a comforting hug from her mother?

The trees at La Branche plantation are particularly odd. One tree is believed to be haunted and another the site of buried treasure. Blossom, who grew up in nearby Kenner, remembers when she was a child that people used to dig under the live oak trees on the property in hopes of finding buried treasure. An uncle of Salvador led a digging expedition one night to find the fortune, but a fierce looking man appeared shaking his finger as if to warn the uncle not to proceed. The poor person digging in the hole leaped out as fast as he could and the whole crew fled, abandoning their pursuit. It is usually prudent not to upset a ghost. This treasure is rumored to be Jean Lafitte's; perhaps he or one of his men still protects his loot. After all it is known that pirates always killed one of their own to have a permanent ghostly guard for the buried stash.

On several occasions members of the family have found smoke inexplicably coming out of the ground when they dig holes, especially near one of the haunted live oak trees. Once, Blossom's daughter Karen and an employee were digging a posthole by a tree when smoke began pluming out of the ground. The worker refused to continue, as he was so frightened. Lisa and Blossom once saw smoke coming out of the bottom of their prized pecan tree. While they did not see any sign of a fire, the smoke made them fear that the largest pecan tree in the state of Louisiana would be reduced to ash and cinders. Instead the following day the tree was untouched, with no trace of damage from the spectral smoke of the night before. Salvador also insists that there is a large iron box located under the pecan tree, which has been detected through probing the ground with metal rods. Could this be the treasure of Jean Lafitte? The Lentinis are not prepared to cut it down to find out.

A further oddity on this site is the orientation of four live oak trees into a square with a mound in the center. One day a Houma Indian came to the plantation and explained that when there are four trees positioned like that there is an Indian Chief buried in the center. He said there would have been a persimmon tree planted over the body; no tree exists today, but the live oak trees are certainly old. Could the numerous strange occurrences on this property be in some way related to this burial?

Without a doubt there are ghosts at La Branche Plantation Dependency. With Hitler's horse on the property, and perhaps the buried treasure of Jean Lafitte, this plantation has more than its share of sources for hauntings. A peaceful quality envelopes La Branche, but a foreboding sense of the eerie past may touch you as well. This paradoxical feeling is difficult to find elsewhere. So as you wander the River Road, stop at La Branche Plantation. You will have an interesting tour, and who knows, you just might see a ghost or two.

II

Pitot House
A Spirit who Loves Flowers

P itot House is a rare oasis in the city of New Orleans, not far from the bustling French Quarter and a short walk to City Park. Standing on Bayou St. John for over two hundred years, this West Indies style house is an extraordinary reminder of what country life was like surrounding the city of New Orleans at the turn of the nineteenth century. The spirit of Pitot House seems content; she is quiet and makes her presence known infrequently in a very gentle, alluring way: the sweet perfume of flowers.

Pitot House was built between 1799 and 1805 and has stucco-covered, brick-between-post construction and had a double pitched hipped roof. All the interior rooms open to a porch, loggia, or gallery. These exterior spaces comprise nearly as much square footage as the interior and were a significant part of the living space of the home. With huge windows, large opposing doors, and no interior hallways, the house was designed to take advantage of cross-breezes. Visitors experience the effectiveness of these windows and doors first hand as cooling breezes off Bayou St. John make the house pleasant and comfortable even during the hot Louisiana summers.

The builder of the house may have been Hilaire Boutté, a prominent builder in New Orleans at the time. We do know the house was built for Bartholemé Bosque, a leading Spanish merchant and ship owner. Bosque's ownership did not last long as he sold the property on May 28, 1800 to Joseph Reynes. Reynes completed the house and added property at the back. The plantation was sold again in 1805 to Madame Marie Tronquet, who continued the work on the house, possibly adding the porch on the south side and several outbuildings. In 1810, she sold the property for $8,400 to James Pitot, which included the house, outbuildings, and thirty acres, which today would encompass seven city blocks.

James Pitot lived in the house until 1819 when it passed to Albin Michel and then to Felix Ducayet in 1848. Ducayet altered the house dramatically including changing the distinctive double-pitched roof to a hipped roof and adding two dormer windows in the front and one in the rear. The Tissot family then owned the property until 1904 when it finally passed to Mother Francis Xavier Cabrini who bought the property for the

Missionary Sisters of the Sacred Heart.

Pitot House was moved from its original location. The Missionary Sisters of the Sacred Heart planned on building a high school at the site in the 1960s. The Louisiana Landmarks Society intervened and saved Pitot House from demolition, with the caveat that the house be relocated a few hundred feet to the site of the Desmare Playground. Painstakingly the house was dismantled, transported, and reassembled as best as possible, although the solid brick walls of the ground floor could not be moved and had to be re-built. Hurricane Betsy further damaged the house in 1965 but the Louisiana Landmarks Society carefully restored the building, furnishing it with American and Louisiana antiques such as would have been in the house during Pitot's occupation.

Pitot House

James Pitot was the house's most famous occupant, the first elected mayor of the city of New Orleans in 1804. James was not actually Pitot's original name. He was born Jacques, and from 1782 to 1792 supervised trade between coffee, indigo and sugar planters in Saint Dominigue (now Haiti) to their customers in Europe and the United States. But when the French Revolution and slave rebellions in Saint Dominigue ended the planters' livelihoods, Pitot fled to America. He settled first in Philadelphia where he changed his name and married Marie-Jeanne Marty, who also had fled the uprisings in Saint Dominigue. Shortly after their marriage they made their way to New Orleans where they set up home in a Royal Street townhouse and Pitot established a successful import-export business. As mayor he quickly set about making civic improvements. Pitot had the streets paved, improved the hospital, built city schools, opened ferry services to the Mississippi, ordered a census, and established a mounted patrol for unsavory neighborhoods. All of this in a very short time as he left the mayor's office in 1805 to return to his business interests.

When Pitot purchased the home on Bayou St. John he had three children. After the War of 1812, during which the British blockade caused irreversible hardship to his import-export business, Pitot was appointed as Judge of the Parish Court. He held the position until his death on November 4, 1831. Tragedy fell on his family on November 30, 1815 when Marie-Jeanne died giving birth to twin girls. Sadly both girls soon joined the spirit of their mother, one died on December

5, 1815 and the other on January 22, 1816. In July of that year Pitot married Genevieve-Sophie Nicolas and they were blessed with two children. In 1819 continued financial troubles forced Pitot to sell his beloved home on Bayou St. John and he moved with his family to the corner of Bourbon and Gov. Nicholls Streets.

Judge Pitot left the house; however, it is uncertain the first Mrs. Pitot ever did. Perhaps she causes the strong floral scent that permeates the master bedroom and the dining room. The bouquet has been so strong and distinct that visitors have asked Director Myrna Bergeron what kind of potpourri is used in the house. The answer of course is that none is used, nor any cleaning agents with floral scents. Staff members also say they have sensed the perfume of flowers where none were found.

Another possibility exists as the source of this pleasant fragrance. Mother Cabrini lived in the house for a short while after purchasing it and is known to have loved flowers. We can imagine her shadowy form floating on Esplanade Avenue, passing onto Moss Street, and drifting through the walls into Pitot House, bringing a ghostly bouquet of her beloved flowers from beyond.

A psychic visiting the house confirmed the presence of a female spirit, although her identity was not established. During the psychic's investigation she became visibly affected in the Blue Room, or master bedroom. She was unable to breath, became red in the face, and needed to leave the room. Perhaps the resident spirit did not wish to be probed and poked by a mortal, but simply to be left alone, quiet, luxuriating in flowers.

The spirit that haunts Pitot House has been especially quiet lately. Perhaps she is content now that the house has been saved and is in good hands. Or maybe she is simply resting peacefully for a time in another dimension. If you visit Pitot House you will be in for an architectural treat in the heart of New Orleans. As the only plantation home in New Orleans open to the public Pitot House provides a glimpse back in time that few houses in the city can match. And although the spirit has been resting for some time, it is possible that you might catch the sweet scent of flowers. If you do, remember no natural source for the aroma has ever been found. You may have been welcomed by Mrs. Pitot herself, who feels you are a kindred soul. But don't worry, the spirit is quiet, doesn't like to show herself, seems content, and loves flowers.

12

Woodland Plantation
A Boy in the Night

In Louisiana's deep Delta country one survivor remains of the bygone plantation life. As you drive south from New Orleans towards Woodland Plantation the landscape changes; groves of orange and satsuma trees replace the sugar cane fields still evident around River Road plantations between New Orleans and Baton Rouge. Approaching Woodland Plantation you begin to feel as if you are reaching the edge of the earth. The narrow strip of land jutting out into the gulf is both the most eastern and southern part of the state. Today it is known as a "Sportsman's Paradise," full of year round opportunities for fishing and gaming. Its past is seedier, replete with stories of the pirate Jean Lafitte and a pair of boat captains engaged in illegal slave trade. While Woodland Plantation might be the only remaining plantation on the West side of the Mississippi River in the deep Delta country, more than its structures endure as a reminder of its past. A congregation of spirits continues to call this plantation home.

William Johnson and George Bradish were sea captains originally from Nova Scotia, who had an unusual domestic arrangement. The men and their wives established Magnolia Plantation as a joint domicile along the West bank of the Mississippi in the Louisiana Delta. While seemingly successful at first, this living situation deteriorated as the families grew older and thoughts of inheritance to their children loomed large. Certainly such a unique arrangement was also trying on their wives who began to fight bitterly as the years passed. In order to have a property to leave to his sons and to please his wife, William Johnson sold his share of Magnolia Plantation to George Bradish and set about planning his own plantation home. From years of working throughout the Delta, William Johnson had noted a plot of land that remained safe and dry despite the harsh and unpredictable climate of Louisiana's Delta coast. It was here that he built Woodland Plantation, a decision that appears to have been a wise one, for Woodland survived long after Magnolia faded away.

Woodland Plantation was built in 1834 as a one and a half story raised four thousand square foot Creole cottage. It has five dormer windows on each façade and Greek Revival features, such as shoulder moldings over the French doors. The front and rear of the house are the same except for minor architectural details. The front porch is twelve feet deep and the back porch is ten feet deep. Typical of plantation homes, the house originally faced the river and what is now perceived to be the front of the home is actually the back. The plantation property itself was twenty-one hundred acres planted in sugar cane.

Woodland Plantation

The house has a basic central hall plan, with the hall extending from front to rear; however, the hall is off-center, a feature usually found in homes with later additions. Today the dining room has a gleaming mahogany dining table and the fireplace in this room has been restored to working order. The entry foyer is an inviting space with an impressive three-tier iron chandelier. The upstairs of the house has a large hall furnished as a sitting area around which there are several bedrooms. Overnight guests can stay in one of the nine bed and breakfast rooms on either the first or second floors.

Originally the property featured four large brick slave quarters. These quarters were heavily damaged by Hurricane Betsy in 1965, and the remaining bricks were apparently looted, for nothing remains today. It is believed that these slave quarters were in place prior to the building of Woodland Plantation and that William Johnson and George Bradish used them to conduct an illegal slave trade with the pirate Jean Lafitte. Jean Lafitte would raid ships in the Gulf of Mexico and bring the stolen slaves up Grand Bayou and store them in the slave quarters. Grand Bayou was a shortcut to the Gulf from the slave quarters since the Gulf was located only twenty-five miles away on the bayou but seventy-five miles away on the Mississippi River. William Johnson and George Bradish would then travel from Magnolia Plantation to pick up the slaves and trade them downriver.

At the site of the four brick slave quarters, an old church, long ago deconsecrated, has been given a new life. Spirits of Woodland Hall is a Gothic chapel, circa 1880, which was moved to Woodland Plantation in 1998. This hall has been restored and now is used for dinners, weddings, receptions, and conferences.

William Johnson's son George Johnson managed the plantation until 1856 when it was taken over by William's other son, Bradish Johnson, in whose hands it remained until it was sold to Theodore Wilkinson in 1897. Theodore's failed bid for governor cost him financially and he was forced to sell the property to his brother Horace, who lived in the house until his death in 1941. Although Woodland remained in the Wilkinson family, only caretakers lived in the home after Horace's death. The remaining thirteen heirs fought bitterly over the inheritance of the property, delaying restoration. During this fifty-six year legal dispute the house suffered greatly. An auction of the property was ordered by Judge William Roe in order to satisfy the claims of all the heirs. Jacques and Claire Creppel and their son Foster won the auction and acquired Woodland in 1997. Today the Creppel family has restored the house to its former grandeur, including painting rooms in their original colors. This restoration was a labor of love that almost didn't occur due to some unlikely interference: the plantation ghosts.

When Foster Creppel arrived at Woodland Plantation he was more interested in the surrounding land and what it would offer for a prospective eco-tourism business. Historic preservation of a plantation home was the furthest thing from his mind. The house appeared beyond repair with owls living in the attic, broken dormer windows, and junk cluttering the second floor. Still Foster moved in and began restoration planning.

However, every night for three months Foster Creppel's sleep was disturbed by loud footsteps and the sounds of furniture being moved on the second floor. Night after night this happened, not allowing him a restful sleep. Every time Foster searched for intruders, there was never any one there. And the noises did not seem like owls or creaking old windows. Foster says, "It was like there were several people up there." One night at three o'clock in the morning the television, on a stand directly outside his room, came crashing to the floor. Armed with his shotgun and prepared for an intruder, Foster burst out of his room, only to find himself alone, with the television at his feet. After three months of nightly awakenings Foster had enough of his supernatural housemates and in his loudest voice proclaimed an ultimatum. "If you wake me up one more time I am going to burn this whole house down." Apparently this was enough to encourage the ghosts to quiet down. Foster never again heard the ghostly residents of Woodland Plantation.

Is Woodland Plantation still haunted? Or did this threat ward off the ghosts for good? It appears that while the ghosts have made their peace with the owner of the site, they are still willing to annoy overnight guests. On numerous occasions there are reports at breakfast of unnatural occurrences in the house the previous night.

One gentleman lived in the house as a caretaker for about three years before the Creppels bought the property. One very hot day in July he was cleaning the guest room called The Suite when he suddenly felt a cold breeze on his back and shoulders. He turned and saw three diaphanous people, a man and two women, hovering ominously over him. He fled the house and did not return for three months.

Richard Fern was a guest at Woodland Plantation, staying in the room called The Roost. He had spent the day fishing with Foster and they had a drink before retiring at 1:00 a.m. He describes the night as moonlit and a soft glow came through the curtains into his room. Suddenly at 3:00 a.m. he sat bolt upright in his bed. Although normally a light sleeper Richard was initially unsure of what caused him to wake up with such urgency. The answer came when he looked at the end of the brass bed, where a young boy, about eight years old with fair hair was standing. The boy appeared to be wearing a white shirt, perhaps a nightshirt. Richard could see only the boy's head and shirt poking up over the bed frame. This was very odd and Richard was startled by this unlikely intruder. Rationalizing the apparition, Richard thought surely the boy had become lost from his parent's room. They must have arrived after he retired for the night. There was only one problem with this logic, the door to his room was still locked, from the inside. The boy kept staring at him. He croaked a question: "Who are you?" There was only silence and the infernal eyes of the boy staring persistently at him. This continued for about forty seconds until the boy began to gradually dissolve. When the boy had finally faded away Richard turned on the light and searched the room, but could find nothing to explain his odd visitor.

As a rational person and former police officer, Richard has tried to find an explanation for what occurred that night, but admits it is difficult to explain. Furthermore, he has stayed in the house on two other occasions in two other rooms and nothing occurred, and he knew nothing about the haunting of the house prior to his encounter.

A group of women staying in the house before a wedding were visibly terrified one morning at breakfast. They insisted they were not alone in the house the night before. Apparently they saw and heard many unexplained things, including one of the women having an uninvited visitor in the bathroom. She was taking a shower and while drying herself, she noticed the door had been unlatched from the inside. Was there someone or something lurking in the bathroom? She scanned the room but there was no one. It seemed the bathroom became colder as she shivered in her hopeless search.

Just who the spirits are is unknown. It could be the spirit of William Johnson returning to his plantation home or one of his sons who ran it after him. Or perhaps it is one of the many men who would have stayed as guests at the home. Since Johnson controlled the Delta and all the passenger ships, he had frequent overnight visitors at his home. These guests were often pirates and bandits. Foster feels these men were allowed complete license in the home. Perhaps the footfalls and moving furniture are sounds of ghostly parties and the spirit in the bathroom a lost reveler. The women and the boy are more elusive spirits, and their origin is still unknown.

If there are ghosts at Woodland, they are not destructive or malevolent. They seem to just want a little attention and to enjoy again the earthly warmth of the home where they once stayed. The Creppel's have ambitious plans for the plantation, including a gift shop currently under renovation, a cabana and pool, and a European spa and suites in the overseer's house.

Louisiana's "Sportsman's Paradise" awaits you just beyond the plantation's door. Woodland Plantation has been immortalized on Southern Comfort bottles. The Currier and Ives print of the house was put on the label of this spirit in 1934, the hundredth anniversary of the house, as a true depiction of southern comfort. Travel to Woodland Plantation and you too can experience southern comfort and spirits of the past.

The Cottage Plantation Ruins

13

Burnt Plantations
Lost Spirits

Thus far all the haunted plantations explored in this book have been restored and are open to the public through tours or as a bed and breakfast. This chapter is different. The plantations described over the following pages no longer have visitors roaming the grounds or guests lounging in their rooms. These homes met a fiery end, burning to the ground, never to be inhabited again; never by earthly residents at any rate. But these homes do share one thing in common with the rest of this book: they are all haunted.

"Valcour" Frances Gabriel Aime was the king of the sugar

planters in Louisiana in the first half of the 1800s. His holdings were grander and he was richer than any other planter in the south. He was so wealthy he had his own steamboat to ferry himself and friends to New Orleans at a moments notice. Valcour's plantation was built in the Classical Revival style with extravagant touches of marble mantles, crystal chandeliers, and a Spanish style courtyard. Located on the West bank of the Mississippi, a short distance from Oak Alley, it was entirely self-sustaining. He proved this by winning a $10,000 bet that he could provide for a full dinner party including wine and cigars from his plantation grounds.

Valcour hired a French landscape designer to transform his swampy Louisiana plantation into a fine French estate complete with a grotto, a Chinese pagoda perched on a hill, Roman bridges over an artificial river, a fort, exotic fish ponds, a zoo of exotic animals, and a hothouse filled with tropical and other rare plants imported from around the globe. It was like nothing ever seen before on this side of the Atlantic. The abundance and elaborateness of his gardens earned his home the name "Le Petit Versailles." To ensure the splendor and longevity of the grounds Valcour kept his gardener full-time, an unheard of luxury at the time. It is rumored Valcour hid an estimated two million dollars worth of jewels and money on his plantation grounds for safekeeping during the Civil War. However he never told anyone the location of this hoard and it has never been found.

Valcour and his wife Josephine had four daughters

before the birth of their son, Gabriel. The boy was pampered and spoiled throughout his youth. As a young adult he spent months of the year traveling in Europe instead of learning the business of the plantation. But Valcour humored his only son, waiting patiently for him to settle down.

But the world fell apart for this king of sugar planters who prided himself on besting everyone. His son died in 1854 at the young age of twenty-seven. This shattered the foundations of Valcour's will to live. He lost interest in his sugar plantation, no longer having a son to bequeath his impressive legacy. And then his wife and two of his daughters were taken from him only a few years later. His vigor and strength began to crumble and the plantation mirrored its owner. In 1867 Valcour died and his already neglected plantation fell into further ruin over the years, finally burning to the ground in 1920. Today a small sign on the road marks the location of this once grandest of all plantations. All we now see are fields and a small clustering of trees. Nothing remains of the luxury and opulence of this former plantation.

However, all is not quiet at the Valcour Aime Plantation ruins. Even while the house was neglected and falling into disrepair there were stories of Valcour and his wife Josephine wandering the grounds and peering from the windows of their abandoned home, wondering forever what happened to the legacy they had begun. There are those who say Valcour Aime still wanders the fields and forests on the site of his plantation home. And who can blame him. It must be hard to forget what he had, what he lost, and that in the end nothing remains. Think of the glory of the former plantation as you view the grounds, now given over to the wind, rain, and loneliness. Think of the grief which permeates the air here; it may be Valcour who touches your heart with sadness.

The ruins of the Cottage Plantation survive just south of the city of Baton Rouge. The few columns that still rise proudly out of the ground from the crumbling brick foundations are lonely reminders of the former grandeur of this plantation. Today, there is not even a road sign to tell us the history of this house. Cows graze unceremoniously among the ruins of The Cottage.

The Cottage Plantation was built in 1824 as a wedding gift to Frederick and Frances Conrad from her father, Colonel Abner Duncan. Although named "The Cottage" this was a fine plantation home with twenty-two rooms constructed in the Greek Revival style. Massive rows of columns supported the double galleries. Fanlight windows graced the entrance door and upstairs gallery door. Live oak trees draped with Spanish moss loomed imposingly on the grounds.

Before the Civil War the likes of Zachary Taylor, Henry Clay, and the Marquis of Lafayette enjoyed the hospitality of this plantation. Frances Parkinson Keyes would call this home from 1943-1945 while she penned her famous novel, *The River Road*. The film, *The Band of Angels* starring Clark Gable, was filmed here in 1957. The great days of the house came to a sudden, tragic end on February 18, 1960 when the house was struck by lightening and burnt to the ground. With no way for the firefighters to access water, everyone just watched in dismay as the proud house roared in flames and fell into cinders and ash and oblivion.

But perhaps not all of the plantation's past went up in flames with the house. One persistent ghost may not have been ready to leave his home, despite its destruction.

Mr. Holt was the devoted personal secretary of Frederick Conrad. When Federal troops occupied The Cottage Plantation during the Civil War both Mr. Holt and Mr. Conrad were thrown into jail. Mr. Conrad died during his imprisonment. After the war Mr. Holt was released and he returned to The Cottage. As all the Conrad family had died, he alone cared for the house until his death in 1880. Soon after his death, there were sightings of a ghostly gentleman with white hair and beard poking around the home in his nightshirt. During his lifetime Mr. Holt had stored away everything from bits of cloth to dried biscuits in trunks in the attic. Perhaps this was prudent behavior brought on by the hardships of war. Mr. Holt's ghost was also seen wandering the house with bits of cloth in his hands. There are those who say Mr. Holt still wanders the ruins of his former home, unable to move on, forever trapped by his duty to Mr. Conrad and to The Cottage. If you visit the plantation you might hear the faint sounds of music and laughter. Memories of an earlier, happier time at the plantation before the Civil War still seem to echo among the ruins.

Between Alexandria and Shreveport is the small town of St.

Maurice, just outside of which was the St. Maurice Plantation. Dennis Fort built St. Maurice in 1826. William Prothro acquired the plantation in 1846 and resided there until the Civil War. After the war the house changed hands numerous times before it was restored in 1980. Tragically this home met a fiery end shortly after its extensive restoration.

Prior to the demise of the house there were stories of children, a caretaker, and a lady in the attic who haunted the property. The child ghost would rise every night from the family cemetery and return to the family home. Today the home is mere ruins and vandals damaged the cemetery itself after the home burned. But the child still makes this nightly pilgrimage home. Some say the child ghost was responsible for burning the plantation home to the ground. The real cause of the fire was never determined, but why this ghost child would want to burn his former home is unknown. Another phantom allegedly haunts the location of storied treasure. A previous owner attempted to dig for the treasure in the spot where his metal detector indicated he had struck gold. As he fitfully dug the ground he greedily imagined his bounty, only to find a pickax, perhaps left behind by a successful treasure seeker.

The Viala Plantation was built in 1797 as a raised Acadian cottage by G.P. Viala and was co-owned by his brother-in-law Dumas St. Martin. By some accounts, the pirate Jean Lafitte had ties to this plantation home and visited via Bayou Lafourche. Whether the pirate Jean Lafitte actually stayed at this house is unclear, but we do know that the builder's granddaughter Emma married a Jean Pierre Lafitte in 1845. This was most likely the grandson of the famous pirate. During the 1950s this plantation home fell into disrepair but was given a new life in the 1970s. In 1974 the house was moved downriver, restored, and four years later became the restaurant Lafitte's Landing, operated by world-renown chef John Folse.

Visitors from all over the world enjoyed this first class restaurant in Donaldsonville, near the foot of the Sunshine Bridge on the West bank of the Mississippi River. Some visitors apparently experienced more than fine dining while enjoying a meal. In the October 1985 issue of *Baton Rouge Magazine* Folse admitted that at least six people encountered a ghost's presence while at the restaurant. The spirit seemed to be none other than Emma Viala, who had a difficult life. She turned to drinking in her loneliness over her husband's frequent trips to New Orleans. She became paralyzed from the waist down as a result of a drunken fall down the stairwell in her home, and died when she was only nineteen years old. Those who saw the spirit described a young woman who was paralyzed from the waist down dragging herself across the floor.

Tragically Lafitte's Landing was destroyed by fire on October 25, 1998. The two hundred year old wooden plantation was engulfed in flames in a matter of minutes and there was little that could be done to save the home despite the relentless efforts of firefighters on the scene. Today there is nothing to remind us of the former grandeur of the Viala Plantation.

Lafitte's Landing has moved into Folse's former home, the Bittersweet Plantation in Donaldsonville. Andrew Gingry began building Bittersweet Plantation in 1853 but it was not completed before the Civil War broke out. Gingry himself was shot to death on the rear steps of his home as he exchanged fire with Union soldiers who were raiding his plantation commissary. It was after this tragic event that Gingry's widow began referring to the plantation as "Bittersweet." Today visitors can still enjoy a sumptuous dining experience at Lafitte's Landing at Bittersweet Plantation.

What of the ghost of Emma Viala? Did she fade away along with the home she once haunted? Or did she move away to haunt Folse's new restaurant? It appears the former for all has been quiet at the new Lafitte's Landing and she has not appeared at the site of the Viala Plantation. Perhaps Emma, who was unhappy in life, has finally found some peace in death.

These burnt plantations represent a part of Louisiana's heritage that has been taken from us tragically. Luckily no one was seriously injured in any of these fires. The homes were either vacant or the occupants escaped in time from the smoke, heat, and flames. One would like to think that the ghosts ensured the safe evacuation of the occupants even if they could not save the home itself. We can picture these ghosts terrified in their burning homes. Some may finally have been able to let their pasts go and move on. Others, forever saddened by the loss of their lives and now the loss of their home, still haunt the ruins of these burnt plantations.

Appendix
Plantation Directory

Please contact sites for hours of operation and B&B information.

Chretien Point Plantation
665 Chretien Point Road
Sunset, LA 70584
(337) 662-7050
(800) 880-7050
www.chretienpoint.com
reservations@chretienpoint.com

Destrehan Plantation
13034 River Road
P.O. Box 5
Destrehan, LA 70047
(985) 764-9315
www.destrehanplantation.org
destplan@aol.com

Frogmore Plantation
11054 Hwy 84
Frogmore, LA 71334
(318) 757-2453
www.frogmoreplantation.com
frogmore@bayou.com

La Branche Plantation Dependency House
11244 River Road
St. Rose, LA 70087
(504) 468-8843
www.labrancheplantation.com

Lafitte's Landing Restaurant at Bittersweet Plantation
404 Claiborne Avenue
Donaldsonville, LA 70346
(225) 473-1232
www.jfolse.com
lafittes@eatel.net

Loyd Hall Plantation
292 Loyd Bridge Road
Cheneyville, LA 71325
(318) 776-5641
(800) 240-8135
www.loydhall.com
info@loydhall.com

The Myrtles Plantation
7747 US Hwy 61
P.O. Box 1100
St. Francisville, LA 70775
(225) 635-6277
www.myrtlesplantation.com
myrtles@bsf.net

Oak Alley Plantation
3645 Hwy 18
Vacherie, LA 70090
(225) 265-2151
(800) 442-5539
www.oakalleyplantation.com
contactus@oakalley.com

Ormond Plantation
13786 River Road
Destrehan, LA 70047
(985) 764-8544
www.plantation.com
ormond@accesscom.net

The Pitot House
1440 Moss St.
New Orleans, LA 70119
(504) 482-0312
www.bellsouthpwp.net/l/a/lalndmrk/
lalnmrk@bellsouth.net

Rosedown Plantation
12501 La Hwy 10
St. Francisville, LA 70775
(225) 635-1867
(888) 376-1867
www.crt.state.la.us/crt/parks/rosedown/rosedown.htm

San Francisco Plantation
2646 River Road
P.O. Box 950
Garyville, Louisiana 70051-0950
(985) 535-5450
(888) 322-1756
www.sanfranciscoplantation.org

Woodland Plantation
21997 Highway 23
West Point a La Hache, LA 70083
(504) 656-9909
(800) 231-1514
www.woodlandplantation.com
spiritsofwood@cs.com

Bibliography

Amort, Joanne. *Oak Alley Plantation*. (n.p.: Oak Alley
　　Foundation, 2000.)
Boyer, Marcel. *Plantations by the River*. Baton Rouge:
　　Louisiana State University, 2001.
Brown, Alan. *Shadows and Cypress: Southern Ghost Stories*.
　　Jackson: University Press of Mississippi, 2000.
Calhoun, Nancy Harris and James Calhoun, Eds. *Plantation
　　Homes of Louisiana*. Gretna, LA: Pelican Publishing
　　Company, 1977.
Dickinson, Joy. *Haunted City: An Unauthorized Guide to the
　　Magical, Magnificent New Orleans of Anne Rice*. New
　　York: Carol Publishing Group, 1995.
Edmonds, David C. *Yankee Autumn in Acadiana: A Narrative
　　of the Great Texas Overland Expedition through
　　Southwestern Louisiana October-December, 1863*.
　　Lafayette, LA: The Acadiana Press, 1979.
Gleason, David King. *Plantation Homes of Louisiana and the
　　Natchez Area*. Baton Rouge: Louisiana State
　　University Press, 1982.

Gore, Laura Locoul. *Memories of the Old Plantation Home & A Creole Family Album*. Vacherie, LA: The Zoe Company, Inc., 2000.

Hauck, Dennis William. *Haunted Places: The National Directory*. New York: Penguin Books, 2002.

Kane, Harnett. *Deep Delta Country*. New York: Duell, Sloan & Pearce, 1944.

___. *Plantation Parade: The Grand Manner in Louisiana*. New York: Bonanza Books, 1945.

Klein, Victor C. *New Orleans Ghosts*. Metairie, LA: Lycanthrope Press, 1996.

___. *New Orleans Ghosts II*. Metairie, LA: Lycanthrope Press, 1999.

Kingsley, Karen. *Buildings of Louisiana*. Oxford: Oxford University Press, 2003.

Lane, Mills. *Architecture of the Old South: Louisiana*. New York: Abbeville Press, 1990.

Laughlin, Clarence John. *Ghosts Along the Mississippi: An Essay in the Poetic Interpretation of Louisiana's Plantation Architecture*. New York: Bonanza Books, 1961.

Levatino, Madeline. *Past Masters: The History and Hauntings of Destrehan Plantation*. Destrehan, LA: Levatino & Barraco, 1991.

McAlester, Virginia and Lee. *A Field Guide to American Houses*. New York: Alfred A. Knopf, 2000.

Malone, Lee. *The Majesty of the River Road*. Gretna, LA: Pelican Publishing Company, 1998.

Mead, Robin. *Haunted Hotels: A Guide to American and Canadian Inns and Their Ghosts*. Nashville: Rutledge Hill Press, 1995.

Miller, John Edward. *Treasure in Louisiana: A Treasure Hunter's Guide to the Bayou State.* Kearney, NE: Morris Publishing, 1996.

Montz, Larry and Deana Smoller. *ISPR Investigates: The Ghosts of New Orleans.* Atglen, PA: Whiteford Press, 2000.

Muse, Vance. *Great American Homes: Old New Orleans.* Birmingham, AL: Oxmoor House, Inc., 1988.

Norman, Michael and Beth Scott. *Haunted America.* New York: Tom Doherty Associates, 1994.

___. *Haunted Heritage.* New York: Tom Doherty Associates, LLC, 2002.

___. *Historic Haunted America.* New York: Tom Doherty Associates, 1995.

Poesche, Jessie and Barbara Sorelle Bacot. *Louisiana Buildings 1720-1740.* Baton Rouge: Louisiana State University Press, 1997.

Saxon, Lyle, Edward Dreyer, and Robert Tallant. *Gumbo Ya-Ya.* Gretna, LA: Pelican Publishing Company, 1991.

___. *Lafitte the Pirate.* Gretna, LA: Pelican Publishing Company, 1989.

Sillery, Barbara. *The Haunting of Louisiana.* Gretna, LA: Pelican Publishing Company, 2001.

Smith, J. Frazer. *Plantation Homes and Mansions of the Old South.* New York: Dover Publications, Inc., 1993.

Smith, Kalila Katherina. *Journey Into Darkness: Ghosts and Vampires of New Orleans.* New Orleans: De Simonin Publications, 1998.

Sternberg, Mary Ann. *Along the River Road: Past and Present on Louisiana's Historic Byway.* Baton Rouge: Louisiana State University Press, 2001.

Taylor, Joe Gray. *Louisiana: A Bicentennial History*. New York: W.W. Norton & Company, Inc., 1976.

Taylor, Troy. *Haunted New Orleans: Ghosts and Hauntings of the Crescent City*. Alton, IL: Whitechapel Productions Press, 2000.

Vivano, Christy, L. *Haunted Louisiana: True Tales of Ghosts and other Unearthly Creatures*. Metairie, LA: Tree House Press, 1992.

Wlodarski, Robert and Anne Powell Wlodarski. *Southern Fried Spirits: A Guide to Haunted Plantations, Inns, and Taverns*. Plano, TX: Republic of Texas Press, 2000.

Word, Christine. *Ghosts Along the Bayou: Tales of Haunting in Southwestern Louisiana*. Lafayette, LA: Acadiana Press, 1988.

Index

About The Author

Jill Pascoe is a native of Winnipeg, Manitoba, Canada. She has a B.A. in Archaeology and History from McGill University and a M.A. in Museum Studies from University College London. She has worked at historic houses and museums in Canada, England, New Jersey, and Virginia. This is her first book.

Jill lives in Baton Rouge with her husband Josh, and their cats Lancelot and Guinevere. She enjoys to read, write, and explore haunted locations around the world.

If you have a ghost story to share from any part of the globe please send it to: stories@irongatepress.com.

Order Form

YES! I want_____ copies of Louisiana's Haunted Plantations for $16.00 each.

Name:_____

Address:_____

City:_____

State/Zip:_____

Telephone:_____

email address:_____

Please include 9% sales tax for orders shipped to Louisiana. Canadian orders must be in US funds.

Include $4.00 shipping and handling for one book, each additional book is $2.00.

All orders will be shipped upon receipt.

Please make checks or money orders payable to Irongate Press and return to:

IRONGATE PRESS
P.O. Box 84602
Baton Rouge, LA 70884-4602

Visit us on-line at www.irongatepress.com

Thank you for your order!